Series/Number 07-109

BASIC MATH FOR SOCIAL SCIENTISTS
Problems and Solutions

TIMOTHY M. HAGLE
The University of Iowa

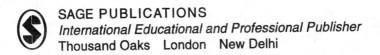

SAGE PUBLICATIONS
International Educational and Professional Publisher
Thousand Oaks   London   New Delhi

For information address:

SAGE Publications, Inc.
2455 Teller Road
Thousand Oaks, California 91320
E-mail: order@sagepub.com

SAGE Publications Ltd.
6 Bonhill Street
London EC2A 4PU
United Kingdom

SAGE Publications India Pvt. Ltd.
M-32 Market
Greater Kailash I
New Delhi 110 048 India

Printed in the United States of America

Library of Congress Cataloging-in-Publication Data

Hagle, Timothy M.
    Basic math for social scientists: Problems and solutions /
author, Timothy M. Hagle.
        p.   cm. — (Sage university papers series. Quantitative
applications in the social sciences; no. 07-109)
      ISBN 0-8039-7285-7 (acid-free paper)
      1. Mathematics—Problems, exercises, etc. 2. Social sciences—
Mathematics—Problems, exercises, etc.   I. Title.   II. Series.
QA43.H245   1996
515′.14′076—dc20

95-41810

96 97 98 99 10 9 8 7 6 5 4 3 2 1

Sage Production Editor: Gillian Dickens

When citing a university paper, please use the proper form. Remember to cite the current Sage University Paper series title and include the paper number. One of the following formats can be adapted (depending on the style manual used):

(1) HAGLE, T. M. (1995) *Basic Math for Social Scientists: Concepts.* Sage University Paper series on Quantitative Applications in the Social Sciences, 07-108. Thousand Oaks, CA: Sage.

OR

(2) Hagle, T. M. (1995). *Basic math for social scientists: Concepts* (Sage University Paper series on Quantitative Applications in the Social Sciences, series no. 07-108). Thousand Oaks, CA: Sage.

CONTENTS

SERIES EDITOR'S INTRODUCTION

The QASS series has now published more than 100 monographs, each explicating some relevant tool for quantitative social scientists. Although the volumes are self-contained and meant to stand alone, certain pairs of them treat a common theme and are better read together. The first example is *Introduction to Factor Analysis* (No. 13) and *Factor Analysis* (No. 14) by Kim and Mueller. Then there is *Confirmatory Factor Analysis* (No. 33) and *Covariance Structure Models* (No. 34) by Long. More recently, we have *Nonparametric Statistics* (No. 90) and *Nonparametric Measures of Association* (No. 91) by Gibbons as well as *Univariate Tests for Time Series Models* (No. 100) and *Multivariate Tests for Time Series Models* (No. 101) by Cromwell and colleagues. This volume is the second in our newest pairing. In his first volume, *Basic Math for Social Scientists: Concepts* (No. 108), Dr. Hagle explained fundamental mathematics concepts behind data analysis techniques. In this second volume, he returns to the same concepts, now teaching them by working through problems.

The goal here is "learning by doing." Readers actually apply the concepts they have studied in the earlier volume (or elsewhere) to problem solving. Chapter 1 introduces an algebra review, sets, permutations, and combinations. Chapter 2 deals with limits and continuity, Chapter 3 differential calculus, Chapter 4 multivariate functions and partial derivatives, Chapter 5 integral calculus, and Chapter 6 matrix algebra. The chapters have a consistent, pedagogically strong style. For each topic, the rules are reviewed, illustrative problems are given to show the operations, and then homework is assigned. Fortunately, the homework problems are numerous. For instance, Chapter 1 offers 44. Furthermore, in Appendix B, the answers are provided, so students can check their work.

The problems are basically of two kinds. First, the traditional, abstract math problems teach mastery of technique. Then, the "story" problems teach application of technique to real situations. Dr. Hagle's set of "story" problems is special, including many social science illustrations—for example, Venn diagrams of voters, tree diagrams of sociology research assistants, choice alternatives in a political action group, rate of change for education test scores, the optimal number of items in a survey, production efficiency in a factory, votes from campaign advertising, budget allocation, population projection, unemployment probability, sample size determination, and utility maximization.

Because the explication is tailored to social science students, appreciation of the material is heightened in comparison to "straight" college math class treatments. Further, Dr. Hagle's informal—and informative—writing draws in the reader. For instance, in evaluating a limit, he observes that the result "looks messy" but goes on to explain how it is now "much easier to evaluate." In finding a second derivative, he also notes that "I could have used different letters for the functions" but the problem would become "more confusing," and he advises students to "be careful when integrating variables with negative exponents." He then tells why. There are many examples like this throughout. Dr. Hagle has taught this material repeatedly to social science graduate students and knows just where they need that extra help. This volume, along with its companion, are without peer, either in terms of quality or price, for equipping social scientists with their basic math needs.

—*Michael S. Lewis-Beck*
Series Editor

PREFACE

This monograph is a companion to *Basic Math for Social Scientists: Concepts.* The first monograph is an introduction (or refresher) to many of the mathematical concepts and techniques that underlie quantitative analysis in the social sciences. Because the first monograph covered a great deal of material, there was little room to present more than a few examples. The purpose of this monograph is to provide more worked-out examples to illustrate the concepts contained in the first monograph. In addition, this monograph contains problems sets that may be used as assignments or for additional practice. The answers to the problems sets are in Appendix B.

This monograph is divided into the same chapters and primary sections as *Concepts,* with problems sets added at the end of each chapter. Although this monograph may be used independently of the first one, readers should be familiar with the concepts contained in the first monograph (and some sections rely on *Concepts* for explanations of notation and usage) to do the problems. In addition, I occasionally will present alternative methods or techniques for which there was not room in the first monograph.

As with the first monograph, my approach to the material is informal and there will be no proofs. Those wishing a deeper understanding of the material can turn to more complete treatments of these topics contained in mathematics textbooks that are available in any college bookstore.

MATH SYMBOLS AND EXPRESSIONS

∃	There exists
∀	For all, for any
∴	Therefore
⇒	Implies
∍	Such that
|	Such that
→	Goes to, approaches
Δ	Delta, the change in
iff	If and only if
∈	Is an element of
∉	Is not an element of
≈	Nearly equal to
≅	Equals approximately
≡	Is equivalent to
≠	Not equal to
×	Multiplied by
[ ]	Matrix, closed interval
| |	Absolute value, determinant
( )	Parentheses, open interval, point, matrix, combinations, etc.
{ }	Set
Σ	Summation
Π	Product
∂	Partial differential
∫	Integral sign
!	Factorial sign
ln	Natural logarithm
$\log_b$	Logarithm to the base b
e	Base (≈ 2.718) of natural logarithms
π	Pi, ≈ 3.1416
μ	Mu, mean
σ	Sigma, standard deviation
σ^2	Sigma squared, variance
ℜ	Real numbers
∞	Infinity
∅	The empty set
∪	Union sign
∩	Intersection sign
⊆	Is a subset of
⊂	Is a proper subset of
lim	Limit of a function
∘	Composition of two functions (small circle)

BASIC MATH FOR SOCIAL SCIENTISTS
Problems and Solutions

TIMOTHY M. HAGLE
University of Iowa

1. INTRODUCTION

1.1 Algebra Review

Recall the rules for exponents:

1. $a = a^1$; if the exponent is 1, it usually is assumed and not written

2. $a^m \times a^n = a^{m+n}$

3. $(a^m)^n = a^{m \times n}$

4. $(a \times b)^n = a^n \times b^n$

5. $\left(\dfrac{a}{b}\right)^n = \dfrac{a^n}{b^n}$, for $b \neq 0$

6. $\dfrac{1}{a^n} = a^{-n}$

7. $\dfrac{a^m}{a^n} = a^{m-n} = \dfrac{1}{a^{n-m}}$

8. $a^{1/2} = \sqrt{a}$, because (from #2) $a^{\frac{1}{2}} \times a^{\frac{1}{2}} = a^{\frac{1}{2}+\frac{1}{2}} = a^1 = a$

9. $a^{1/n} = \sqrt[n]{a}$

10. $a^{m/n} = (a^{1/n})^m = (a^m)^{1/n} = \sqrt[n]{a^m} = (\sqrt[n]{a})^m$

11. $a^0 = 1$, because $a^0 = a^{n-n} = (a^n)(a^{-n}) = \dfrac{a^n}{a^n} = 1$

12. 0^0 is undefined (although it may be defined specially for a given context)

Problem 1.1.1. Use the rules of exponents to evaluate the following expressions.

a. $6^3 = 6 \times 6 \times 6 = 216$: This illustrates the basic meaning of an exponent.

b. $(3 \times 4)^2 = (3 \times 4) \times (3 \times 4) = 3^2 \times 4^2 = 9 \times 16 = 144 = 12^2$: This is an application of rule 4. The last step demonstrates that we could have obtained the same result by first multiplying the numbers inside the parentheses and then squaring the product.

c. $(5^2)^2 = 5^{2 \times 2} = 5^4 = 625 = 25^2$: This is an application of rule 3. Again, the last step shows that we could have obtained the same result by first squaring the 5 inside the parentheses, then squaring 25.

d. $3^{-4} = \dfrac{1}{3^4} = \dfrac{1}{81}$: This is an application of rule 6, which shows that a number raised to a *negative* exponent is equal to 1 over that number raised to the positive value of the exponent.

e. $\left(\dfrac{1}{16}\right)^{1/4} = \left(\dfrac{1^{1/4}}{16^{1/4}}\right) = \left(\dfrac{1^{1/4}}{(2^4)^{1/4}}\right) = \dfrac{1}{2^{4 \times (1/4)}} = \dfrac{1}{2^{4/4}} = \dfrac{1}{2}$: There are three basic steps

to solving this problem. First, apply rule 5 to see that

$$\left(\dfrac{1}{16}\right)^{1/4} = \left(\dfrac{1^{1/4}}{16^{1/4}}\right)$$

Second, convert the number 16 to its exponent form of 2^4. Finally, apply rule 10 to see that $(2^4)^{1/4} = 2^{4 \times (1/4)} = 2^{4/4} = 2$.

f. $\left(\dfrac{8}{27}\right)^{2/3} = \dfrac{8^{2/3}}{27^{2/3}} = \dfrac{(2^3)^{2/3}}{(3^3)^{2/3}} = \dfrac{2^{3 \times (2/3)}}{3^{3 \times (2/3)}} = \dfrac{2^2}{3^2} = \dfrac{4}{9}$: For this problem you also must convert the numerator to exponent form.

g. $(0.01)^{-2} = \left(\dfrac{1}{100}\right)^{-2} = \dfrac{1^{-2}}{100^{-2}} = \dfrac{1}{(10^2)^{-2}} = \dfrac{1}{10^{2 \times (-2)}} = \dfrac{1}{10^{-4}} = 10^4 = 10{,}000$:

There are several ways to approach this problem. I chose to convert .01 to its fraction form and applied rule 5 to the result. I then converted the denominator to exponent form, applied rule 6, and finished the problem by raising 10 to the fourth power.

h. $(a^2 b^{-1} c^{3/5})(a^{-3} b^{1/2} d^4) = a^{2-3} b^{-1+1/2} c^{3/5} d^4 = a^{-1} b^{-1/2} c^{3/5} d^4$: This is an application of rule 2 using the variables a, b, c, and d. One might want to further simplify the final expression by moving the a and b terms to the denominator to remove the negative exponents (i.e., $\dfrac{c^{3/5}d^4}{ab^{1/2}}$).

i. $(a + b)^2 = (a + b)(a + b) = a^2 + ab + ab + b^2 = a^2 + 2ab + b^2$: Like part (a), this is a demonstration of the basic meaning of an exponent. It shows that the rules of exponents apply to unknown quantities as well as known numbers. It is a good idea to be systematic about initially multiplying each term of one factor with each term of the other(s). Once all the terms have been multiplied, you can combine like terms. ("Like" terms have the same variables to the same powers, such as $ab + ab = 2ab$, but not a^2b and ab^2.)

Recall the rules for logarithms.

1. $\log (a \times b) = \log a + \log b$

2. $\log (a/b) = \log a - \log b$

3. $\log (a^n) = n \log a$

4. $\log_a a^x = x$; $\ln e^x = x$

5. $a^{\log_a x} = x$; $e^{\ln x} = x$

Problem 1.1.2. Evaluate the following.

a. $\log_{10} 100 = \log_{10} 10^2 = 2 \log_{10} 10^1 = 2 \times 1 = 2$: This is an application of rule 3 showing every step in the process. This problem contains the subscript 10, but recall that we usually leave the subscript off if the base is 10. (Also recall, however, that some disciplines and software packages use *log* to represent the natural log rather than *ln*, so remember to look for a declaration of the base.)

b. $\log 100 + \log 400 = \log 40{,}000 = \log (10{,}000 \times 4) = \log 10{,}000 + \log 4$
    $= \log 10^4 + \log 4 \approx 4 + 0.602 \approx 4.602$: Here we first use rule 1 to combine the two terms. Although we need not have separated the product into two different factors, it shows more clearly how the process works. Of course, we could have approached the problem as:

$$\log 100 + \log 400 = \log 10^2 + \log (4 \times 10^2)$$
$$= \log 10^2 + \log 4 + \log 10^2$$
$$= 2 + 2 + \log 4 \approx 4 + 0.602 \approx 4.602$$

(*Note:* Here, and elsewhere in this monograph, $\approx$ indicates when there has been some "rounding error" in the decimal approximation of a value.)

c. $\log_3 = 81 = \log_3 3^4 = 4$: This is an application of rule 4. Although this problem uses a different base, we are still looking for the solution to the equation $3^x = 81$, where 3 is the base and 81 the number of interest. More generally, the equations $\log_b N = x$ and $b^x = N$ are equivalent.

d. $\ln e = \log_e e^1 = 1$: This problem demonstrates that $\ln \equiv \log_e$ and, using rule 4 again, the log of the base number raised to the first power is 1.

e. $\log 1 = \log 10^0 = 0$: This is another application of rule 4 with 10 as the base and the exponent as 0. Note that we can generalize this result as $\log_a 1 = 0$ because, by definition, $a^0 = 1$ for $a > 0$ (see exponent rule 11).

f. $\log_2 \dfrac{1}{8} = \log_2 2^{-3} = -3$: Here we see that the log of a number can be negative. (Recall, however, that you cannot take the log of a negative number.)

g. $\ln \dfrac{1}{e^{3/2}} = \ln e^{-3/2} = \dfrac{-3}{2}$: Another problem with negative exponents.

Problem 1.1.3. Solve the following.

a. $\log_4 64 = x \Rightarrow \log_4 4^3 = 3 = x$: This is a basic manipulation of the components.

b. $\log x = 3.2 \Rightarrow 10^{3.2} = 1584.9 = x$: This problem is not so neat. To get the above solution you would need to have either a calculator or a log table for base 10. Even without a calculator or log table, however, you should have been able to estimate that the answer was near 1,500. You know that $10^{3.2} = 10^3 \times 10^{1/5}$. Of course, $10^3 = 1,000$, so we just need to estimate $10^{1/5}$. To do so, we can guess that the square root of 10 is a bit more than 3 (actually about 3.16). We are looking for the fifth root of 10, so we might guess that it will be about half the value of the square root or near 1.5. Thus, $1,000 \times 1.5 = 1,500$, which is in the right neighborhood. You certainly would not use such estimates to solve a problem, but they are a good way to determine whether you correctly entered the problem into your calculator. (*Note:* In a problem such as this, you might see the term *antilog*. If $\log_b x = y$, then antilog$_b y = x$, which is another way of saying $b^y = x$.)

c. $\exp 2 = x \Rightarrow e^2 \approx 7.39 = x$: This problem introduces some alternate notation. You may sometimes see $\exp_b x$ used to represent b^x. Unlike *log*, however, if you see *exp* without a subscript, the base generally is assumed to be e. Thus, $\exp 3 = e^3 \approx 20.09$. (Recall that $e \approx 2.718$.)

Problem 1.1.4. Graph the following functions. Begin by selecting a few points for each function.

a. $y = -2x + 3$			b. $y = x^3$	

x	y		x	y
-2	7		-2	-8
-1	5		-1	-1
0	3		0	0
1	1		1	1
2	-1		2	8

These functions are graphed in Figure 1.1. Notice that $y = -2x + 3$ is a linear function, which means that its graph is a straight line. Of course, we only need two points to graph a linear function. The second function, $y = x^3$, is more complex and, as with many complex functions, we need more points to see its shape.

Recall that if you know the coordinates of two points, you can find the equation of the line passing through them using the formula

$$y - y_1 = \frac{y_2 - y_1}{x_2 - x_1}(x - x_1),$$

where (x_1, y_1) and (x_2, y_2) are two points on the line.

Problem 1.1.5. Find the equation of the lines that pass through the following pairs of points.

a. $(-4, 5)$ and $(1, -2)$: Substituting into the above formula we have

$$y - 5 = \frac{-2 - 5}{1 - (-4)}[(x - (-4)] \Rightarrow y - 5 = \frac{-7}{5}(x + 4)$$

$$\Rightarrow y = \frac{-7}{5}(x + 4) + 5 = \frac{-7}{5}x - \frac{28}{5} + \frac{25}{5} = \frac{-7}{5}x - \frac{3}{5}$$

$$\therefore y = \frac{-7}{5}x - \frac{3}{5}$$

After making the appropriate substitutions, we simply manipulate and simplify the equation to get it into the conventional $y = mx + b$ form, where

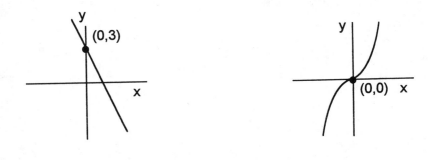

a. $y = -2x + 3$

b. $y = x^3$

Figure 1.1. Graphs for Problem 1.1.4

m is the slope of the line and b is the y-intercept (i.e., where the line crosses the y-axis).

b. (2, 2) and (6, 3): Substituting into the formula we have

$$y - 2 = \frac{3-2}{6-2}(x - 2) \Rightarrow y - 2 = \frac{1}{4}(x - 2)$$

$$\Rightarrow y = \frac{1}{4}x - \frac{2}{4} + 2 = \frac{1}{4}x + \frac{3}{2}$$

$$\therefore y = \frac{1}{4}x + \frac{3}{2}$$

Problem 1.1.6. Find the equations for the following lines.

a. The line with slope $m = \frac{1}{2}$ that passes through (0, 0). If the line passes through (0, 0), we know that the y-intercept is 0. Because we are given the slope, $m = \frac{1}{2}$, the equation for the line must be $y = \frac{1}{2}x + 0$. (You need not indicate that the y-intercept is 0. I included it this time only to show that I did not forget it.)

b. The line with slope $m = \dfrac{-1}{3}$ that passes through $\left(\dfrac{1}{3}, -1\right)$. The equation must satisfy the formula $y = mx + b$ or, because we know that $m = \dfrac{-1}{3}$, $y = \dfrac{-1}{3}x + b$. We also know the line passes through the point $\left(\dfrac{1}{3}, -1\right)$; thus we can substitute these values into the equation and solve for b as follows:

$$y = \frac{-1}{3}x + b \Rightarrow -1 = \frac{-1}{3}\left(\frac{1}{3}\right) + b \Rightarrow -1 = \frac{-1}{9} + b \Rightarrow b = \frac{-8}{9}$$

$$\therefore y = \frac{-1}{3}x - \frac{8}{9}$$

Problem 1.1.7. If two lines are not parallel, they can have only one point in common. Find the point of intersection for the following pairs of lines.

a. $y = 3x + 4$ and $y = -2x + 1$: Because the right-hand portion of each equation is equal to y, we can set them equal to each other and solve for x.

$$3x + 4 = -2x + 1 \Rightarrow 5x = -3 \Rightarrow x = \frac{-3}{5}$$

Now substitute this value for x into one of the equations to find the value for y:

$$y = 3x + 4 = 3\left(\frac{-3}{5}\right) + 4 = \frac{-9}{5} + \frac{20}{5} = \frac{11}{5}$$

To verify that the point $\left(\dfrac{-3}{5}, \dfrac{11}{5}\right)$ is also a point on the second line and therefore the point of intersection, enter the values into the second equation:

$$y = -2x + 1 \Rightarrow \frac{11}{5} = -2\left(\frac{-3}{5}\right) + 1 = \frac{6}{5} + \frac{5}{5} = \frac{11}{5}$$

b. $y = 4x$ and $y = -x - 7$: Again, set the right-hand portions equal to each other, solve for x, and then continue as before:

$$4x = -x - 7 \Rightarrow 5x = -7 \Rightarrow x = \frac{-7}{5}$$

$$\therefore y = 4x = 4\left(\frac{-7}{5}\right) = \frac{-28}{5}$$

$$\therefore y = -x - 7 \Rightarrow \frac{-28}{5} = -\left(\frac{-7}{5}\right) - 7 = \frac{7}{5} - \frac{35}{5} = \frac{-28}{5}$$

Problem 1.1.8. Multiply the following pairs of linear factors together to form polynomials.

a. $(x + 4)(x - 3) = x^2 + 4x - 3x - 12 = x^2 + x - 12$

b. $(3x + 1)(2x - 2) = 6x^2 + 2x - 6x - 2 = 6x^2 - 4x - 2$

c. $(x - 7)(-3x + 1) = -3x^2 + 21x + x - 7 = -3x^2 + 22x - 7$

d. $\left(\frac{1}{2}x + 3\right)(-2x + 2) = -x^2 - 6x + x + 6 = -x^2 - 5x + 6$

Problem 1.1.9. Factor the following polynomials into linear components.

a. $x^2 + x - 12 = (x + 4)(x - 3)$: This is, of course, the reverse of part (a) of the previous problem. Because the coefficient of the x^2 term is 1, we need only find two numbers that when multiplied equal -12 and when added equal 1 (which is the coefficient of the x term). It should not take long to discover that 4 and -3 are the numbers we need.

b. $3x^2 - 14x - 5 = (3x + 1)(x - 5)$: This problem is a bit more difficult because the coefficient on the x^2 term is greater than 1. Even so, it should not take long to find the proper combination of factors.

c. $4x^2 + 12x + 8 = 4(x^2 + 3x + 2) = 4(x + 2)(x + 1)$: Here you should see that 4 can be factored from each term. Although not essential, if you do not factor it out you can get two different (but correct) answers $(4x + 8)(x + 1)$ and $(x + 2)(4x + 4)$.

d. $-x^2 - 6x + 27 = -1(x^2 + 6x - 27) = -1(x + 9)(x - 3)$: Again, if you did not factor out the -1, you could get two different (but correct) answers.

Problem 1.1.10. Solve the following equations.

a. $x^2 - 2x - 8 = 0 \Rightarrow (x - 4)(x + 2) = 0$, $\therefore x = -2, 4$: To find when the equation is equal to 0, begin by factoring it into its linear components. Because the two factors are multiplied together, if either is equal to 0, then the entire left side is equal to 0. Thus, we solve the two simpler equations $x - 4 = 0$ and $x + 2 = 0$ to obtain the values for x.

b. $2x^2 - x - 3 = 0 \Rightarrow (2x - 3)(x + 1) = 0$, $\therefore x = -1, \dfrac{3}{2}$: As before, you might need to experiment a bit to find the right combination of factors that produce the original equation. Once found, set each equal to 0 and solve separately.

c. $6x^2 + 8x - 8 = 0$: Although we could continue to find the factors by trial and error, with more complex equations it is easier to use the quadratic equation. Recall that the solution to a polynomial equation of the form $ax^2 + bx + c$ is

$$x = \frac{-b \pm \sqrt{b^2 - 4ac}}{2a}$$

Applying this formula to the above equation we have

$$x = \frac{-8 \pm \sqrt{8^2 - 4 \times 6 \times (-8)}}{2 \times 6} = \frac{-8 \pm \sqrt{64 + 192}}{12} = \frac{-8 \pm \sqrt{256}}{12} = \frac{-8 \pm 16}{12}$$

$$\therefore x = \frac{-8 + 16}{12} = \frac{8}{12} = \frac{2}{3}, x = \frac{-8 - 16}{12} = \frac{-24}{12} = -2$$

d. $x^2 + x + 1 = 0$: Using the quadratic formula we have

$$\frac{-1 \pm \sqrt{1^2 - 4 \times 1 \times 1}}{2 \times 1} = \frac{-1 \pm \sqrt{-3}}{2}$$

The negative number under the radical sign means the equation has no solution in the real numbers (i.e., it has no real roots). This means that there are no values for x such that $y = 0$ and the graph of $y = x^2 + x + 1$ does not cross the x-axis.

Problem 1.1.11. Find the real roots of the following equations.

a. $x^3 - 4x = 0 \Rightarrow x(x^2 - 4) = 0 \Rightarrow x(x + 2)(x - 2) = 0$, $\therefore x = -2, 0, 2$: Because the highest power in this equation is greater than 2, we initially cannot use the quadratic formula. We can factor an x out of each term on the left side, and doing so tells us that 0 must be one of the roots of the equation. To find the other roots, we solve the equation $x^2 - 4 = 0$. We could use the quadratic formula, but if you see that $x^2 - 4$ is the difference of two squares then you know it can be factored into $(x + 2)(x - 2)$, from which we can easily see that −2 and 2 are the remaining roots. (*Note: Difference of two squares* means that if you have an expression of the form $a^2 - b^2$, it can be factored into $(a + b)(a - b)$.)

b. $x^4 - 1 = 0 \Rightarrow (x^2 + 1)(x^2 - 1) = 0 \Rightarrow (x^2 + 1)(x + 1)(x - 1) = 0$: We initially cannot apply the quadratic formula, but we do have the difference of two squares. After factoring, one of the terms is a difference of two squares and can be factored again. From the linear terms $(x + 1)$ and $(x - 1)$ we can see that two of the roots are $x = -1$ and $x = 1$. There does not seem to be an obvious way to factor $x^2 + 1$, so we can use the quadratic formula:

$$x = \frac{-0 \pm \sqrt{0^2 - 4 \times 1 \times 1}}{2 \times 1} = \frac{\pm\sqrt{-4}}{2} = \frac{\pm 2\sqrt{-1}}{2} = \pm\sqrt{-1}.$$

As before, the negative number under the radical sign indicates that this portion of the original equation has no real roots. Thus, the only real solutions to the original equation are −1 and 1.

Problem 1.1.12. Suppose that a housing contractor has a linear cost function consisting of fixed costs of $300,000 per year and building costs of $54,000 per house. What are the contractor's costs if 11 houses are built in one year? Begin by putting the "story" into numerical form. A linear cost function tells us that the contractor's costs are of the form $y = mx + b$, where b is the y-intercept, m is the slope, x is the number of units, and y is the total cost. In this problem, x represents the number of houses built. "Fixed costs" are those that do not depend on the number of houses built. In the linear equation, they are represented by b. The slope is the "per house" costs (i.e., how much the total costs increase for each house built). Thus, $y = 54x + 300$. (Notice that I divided the fixed and per house costs by 1,000 to simplify the computations. When we calculate y using this equation, we will need to remember to multiply it by 1,000 to get the correct dollar amount.) To calculate the costs for a year in which 11 houses are built, enter 11 for x in the equation and finish the computations: $y = (54 \times 11) + 300 = 594 + 300 = 894$. Remembering to multiply the answer by 1,000, we find that the contractor's costs to build 11 houses in a year are $894,000.

Problem 1.1.13. Suppose that the contractor from the previous problem sells each house for $90,000. How many houses must the contractor build to break even? Again, begin by setting up the problem in numerical form. The basic form of the equation is $P = R - C$, where P represents profits, R represents revenues, and C represents costs. To find the break-even point, we must find where $P = R - C = 0$. We know from the previous problem that $C = 54x + 300$. If each house is sold for $90,000, then $R = 90x$ (again dividing by 1,000 to simplify the calculations). Thus,

$$R - C = 90x - (54x + 300) = 36x - 300 = 0 \Rightarrow 36x = 300 \Rightarrow x \approx 8.33$$

Thus, the contractor must build and sell at least 9 houses per year to break even.

1.2 Sets

Problem 1.2.1. Show how the outcomes from flipping a coin three times can be represented on a line or in three dimensions. Figure 1.2a shows the one-dimensional representation of the number of heads that may be obtained from flipping a coin three times. Figure 1.2b shows how the eight possible outcomes of the three flips can be represented so as to retain information on the outcome of each flip by plotting the outcome of the first flip on the x-axis, the outcome of the second flip on the y-axis, and the outcome of the third flip on the z-axis.

Problem 1.2.2. Use set notation to represent the set of outcomes from rolling a pair of dice where the sum of the two dice is equal to 5.

$$S = \{(1, 4), (2, 3), (3, 2), (4, 1)\}$$

This set consists of four elements. Each element is represented by a pair of values in which the first value is the number of dots showing on one die, and the second value is the number of dots showing on the second die. This set contains only some of the possible outcomes from rolling two dice. The set is therefore not a universe, and we use S (for set, subset, or sample) to represent it.

Problem 1.2.3. What do the following expressions in set notation mean? Make a graph for each expression. (Figure 1.3 contains Venn diagrams for each indicated set.)

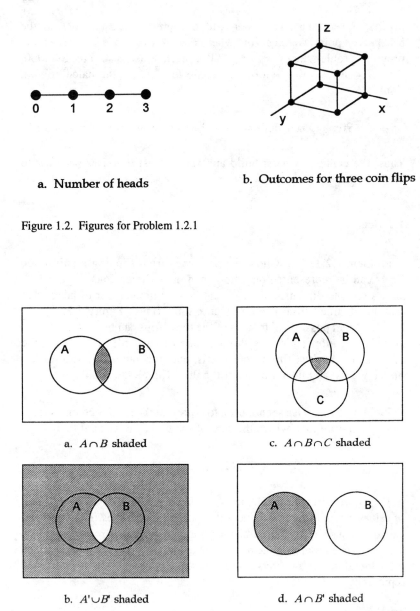

a. Number of heads

b. Outcomes for three coin flips

Figure 1.2. Figures for Problem 1.2.1

a. $A \cap B$ shaded

c. $A \cap B \cap C$ shaded

b. $A' \cup B'$ shaded

d. $A \cap B'$ shaded

Figure 1.3. Venn Diagrams for Problem 1.2.3

a. $A \cap B$: This is the intersection of sets A and B. The resulting set consists of all those points that are elements of set A *and* of set B.

b. $A' \cup B'$: This is the union of all points in the complement of A (i.e., all points of the universe U that are not in A) with all points in the complement of B. (Notice that $A' \cup B' = (A \cap B)'$.)

c. $A \cap B \cap C$: This is the set of points that are elements of all three sets.

d. $A \cap B'$: This is the intersection of A with the complement of B. Notice that the sets A and B do not intersect (i.e., $A \cap B = \varnothing$), so $A \subset B'$ and, thus, $A \cap B' = A$. This shows the importance of how the sets are drawn initially.

Problem 1.2.4. A political survey organization claims that among 400 voters interviewed, 250 regularly gave money to Group X, 150 regularly gave money to Group Y, and 25 gave money to both groups. Use a Venn diagram to determine how many voters did not give money to either group. In Figure 1.4, the left circle represents voters who gave money to Group X, and the right circle represents voters who gave money to Group Y. We are told that 25 voters gave money to both groups. In set notation we can represent this as $X \cap Y = 25$. That 25 voters gave money to both groups tells us that 225 voters gave only to Group X (i.e., $X \cap Y' = 225$), and 125 voters gave money only to Group Y (i.e., $X' \cap Y = 125$). This leaves 25 voters who did not give money to either group (i.e., $(X \cup Y)' = 25$).

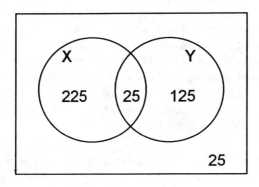

Figure 1.4. Venn Diagram for Problem 1.2.4

1.3 Permutations and Combinations

Problem 1.3.1. Draw tree diagrams for the following.

a. A sociologist has a grant to study professors. One research assistant will be sent to interview each rank of professor (assistant, associate, and full) in the Political Science and Sociology Departments at the Universities of Iowa, Minnesota, and Wisconsin. How many research assistants will be necessary? Figure 1.5a shows that this tree diagram has three levels, one for each decision. After drawing the tree diagram, we can count the endpoints to see that there are 18 endpoints, each of which represents a required research assistant.

b. A political consultant must choose to visit one of three cities in either Michigan, Ohio, or Indiana. Figure 1.5b shows that, given the three states and three possible cities in each state, the consultant has nine possible choices of a city to visit. Notice that unlike the tree diagram for part (a), here it would not make sense to switch the order of the decision: If the choice of a city is made first, there is no longer a choice as to the state.

Problem 1.3.2. How can the sociologist of Problem 1.3.1a determine the number of required research assistants without drawing a tree diagram? Stated another way, the problem is to determine how many ways one can select a university, department, and rank. Because there are three choices for the university, two choices for the department, and three choices for the rank, the answer is $3 \times 2 \times 3 = 18$.

Problem 1.3.3. If a political action group has 20 members, how many ways can a president, vice president, and recording secretary be chosen? If the president is chosen first, there are 20 possibilities. Assuming no member can hold more than one office (i.e., no replacement), there are 19 choices for the vice president. (In a sense, you can think of the choice of vice president as being made from a new set containing only the remaining 19 members.) This leaves 18 members available for the position of recording secretary. Thus, there are $20 \times 19 \times 18 = 6,840$ ways the three officers can be chosen.

Problem 1.3.4. To calculate the number of ways (or permutations) in which r objects can be selected from a group of n objects, we use the formula $P(n, r) = \dfrac{n!}{(n-r)!}$. Use this formula to calculate the following.

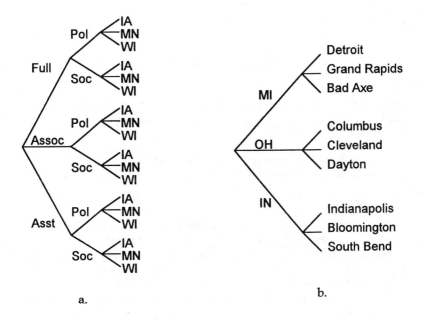

a.   b.

Figure 1.5. Tree Diagrams for Problem 1.3.1

a. $P(10, 3) = \dfrac{10!}{(10-3)!} = \dfrac{10!}{7!} = \dfrac{10 \times 9 \times 8 \times 7 \times 6 \times 5 \times 4 \times 3 \times 2 \times 1}{7 \times 6 \times 5 \times 4 \times 3 \times 2 \times 1}$

$\qquad\qquad = 10 \times 9 \times 8 = 720$

b. $P(8, 5) = \dfrac{8!}{(8-5)!} = \dfrac{8!}{3!} = \dfrac{8 \times 7 \times 6 \times 5 \times 4 \times 3 \times 2 \times 1}{3 \times 2 \times 1}$

$\qquad\qquad = 8 \times 7 \times 6 \times 5 \times 4 = 6{,}720$

c. $P(20, 3) = \dfrac{20!}{(20-3)!} = \dfrac{20!}{17!} = \dfrac{20 \times 19 \times 18 \times 17 \times 16 \times \ldots \times 2 \times 1}{17 \times 16 \times \ldots \times 2 \times 1}$

$\qquad\qquad = 20 \times 19 \times 18 = 6{,}840$

(This, of course, is the more formal treatment of Problem 1.3.3.)

Problem 1.3.5. Permutations assume all n objects in the group are distinct. If some of the objects are not distinct (or if we consider distinct objects to be interchangeable), then there are fewer *different* permutations of the objects. Using the formula

$$\frac{n!}{r_1! \times r_2! \times r_3! \times \ldots r_j!} \; ,$$

where n is a number of objects of which r_1 are the same, r_2 others are the same, . . . , and r_j others are the same, calculate the following.

a. How many different ways can the letters in the word *parrot* be arranged? There are six letters, two of which are the same, so using the formula we have $\dfrac{6!}{2! \times 1! \times 1! \times 1! \times 1!} = \dfrac{6 \times 5 \times 4 \times 3 \times 2}{2} = 6 \times 5 \times 4 \times 3 = 360.$

b. Suppose a presidential candidate wants to make campaign stops in Los Angeles, San Francisco, Portland, Eugene, Seattle, and Tacoma. In how many different ways could the candidate visit the states in which the cities are located? The first two cities are in California, the next two in Oregon, and the last two in Washington. Because we want to know the order of the *states* the candidate could visit, stops in, for example, Los Angeles and San Francisco are the same for purposes of the question. Thus, $\dfrac{6!}{2! \times 2! \times 2!} = \dfrac{6 \times 5 \times 4 \times 3 \times 2}{2 \times 2 \times 2} = 6 \times 5 \times 3 = 90.$

Problem 1.3.6. If we do not care about the order in which the objects are drawn from a group, we use the formula for *combinations*,

$$\binom{n}{r} = \frac{n!}{r! \times (n-r)!} \; ,$$

where n is the number of objects in the group and r is the number of objects to be selected. Using the formula for combinations, calculate the following.

a. Suppose eight candidates are in a primary. In how many ways could the two candidates with the most votes be chosen for a runoff? Entering these values into the formula, we have $\dbinom{8}{2} = \dfrac{8!}{2! \times (8-2)!} = \dfrac{8!}{2! \times 6!} = \dfrac{8 \times 7}{2} = 4 \times 7 = 28.$ Notice that if the order mattered, then we would have $8 \times 7 = 56$ different ways the two top candidates could be listed. Because we do not care about the order, we divide by 2 to obtain the answer 28.

b. A foundation has the money to fund four projects from nine applications. In how many ways can the four projects be chosen? Using the formula for combinations, we have

$$\binom{9}{4} = \frac{9!}{4! \times (9-4)!} = \frac{9!}{4! \times 5!} = \frac{9 \times 8 \times 7 \times 6}{4 \times 3 \times 2} = 9 \times 2 \times 7 = 126.$$

1.4 Homework Problems

1. Evaluate the following.

    a. $(\frac{1}{8})^{4/3}$

    b. $(.0001)^{3/4}$

    c. $(ab^2 c^{-1})(a^3 c^2)$

2. Evaluate the following.

    a. $\log 1000$

    b. $\log_4 256$

    c. $\log_2 (4 \times 8)$

    d. $\log_a a^m$

    e. $\ln (e^{4/3})$

    f. $\ln e^{2/5} + \ln e^{-1/3}$

3. Graph the following relations. Which are functions?

    a. $y = 2x + 1$

    b. $y = 2 - x - x^2$

    c. $2x^2 + y^2 = 1$

4. Find the equation of the lines that pass through the following pairs of points.

    a. $(0, 0)$ and $(1, 2)$

    b. $(3, 2)$ and $(-1, -1)$

5. Find the points where the following pairs of lines intersect.

    a. $y = 2, y = x - 2$

    b. $y = -2x - 1, y = \frac{3x}{2} - 1$

6. Multiply the following.

    a. $(x - 4)(x - 2)$

    b. $(3x - 7)(2x + 4)$

    c. $(-x + 1)(3x + 5)$

    d. $\left(\frac{1}{2}x - 2\right)(2x + 6)$

7. Factor the following polynomials without using the quadratic formula.

    a. $x^2 + 3x + 2$

    b. $3x^2 + 12x + 9$

    c. $-x^2 + 6x - 5$

    d. $x^3 + 2x^2 + x$

8. Find the real roots of the following.

a. $x^2 - 5 = 0$ c. $-5x^3 - 5x = 0$

b. $2x^2 + 7x = 4$

9. Represent the following in set notational form.

   a. The set of outcomes from rolling a pair of dice in which the sum of the dice is equal to 7.

   b. The set of outcomes from rolling a pair of eight-sided dice in which the sum of the dice is equal to 17.

10. Use set notation to describe the shaded area in the Venn diagrams shown in Figure 1.6.

11. Draw tree diagrams for the following.

   a. A "game" consists of the first player rolling a die and the second player flipping a coin. Show the possible ways the game can progress.

   b. A traveler can go to either Florida, Georgia, or Texas by either plane, train, or bus. Show the possibilities.

12. Calculate the following.

   a. $P(13, 4)$ b. $P(31, 2)$

13. Determine the following.

   a. In how many ways can a researcher choose three assistants from six applicants if order matters?

   b. In how many ways can a club choose a president, vice president, secretary, and treasurer if there are 27 members and no one can hold more than one office?

   c. In how many ways can the club in Problem 13b choose its officers if the four past presidents cannot serve as either president or vice president?

14. In how many ways can a student arrange eight textbooks on a shelf? In how many ways can the books be arranged if the student wants to keep the three political science books together, the two sociology books together, and the two psychology books together?

15. Calculate the following.

   a. $\begin{pmatrix} 9 \\ 6 \end{pmatrix}$ c. $\begin{pmatrix} 16 \\ 5 \end{pmatrix}$

   b. $\begin{pmatrix} 11 \\ 7 \end{pmatrix}$

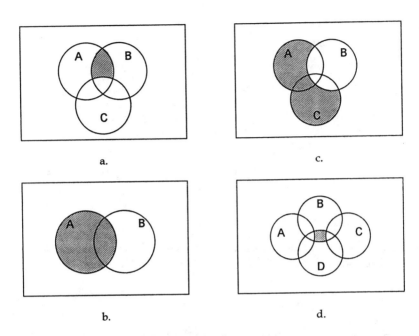

a.

c.

b.

d.

Figure 1.6. Venn Diagrams for Homework Problem 1.10

2. LIMITS AND CONTINUITY

2.1 The Concept of a Limit

Problem 2.1.1. Consider the function $f(x) = x + 1$ on the interval $[0, 2]$. What is $\lim_{x \to 1} (x + 1)$? Use a table of values to show that $f(x) \to 2$ as $x \to 1$.

Begin by constructing a set of data points as follows:

x	$f(x)$		x	$f(x)$
0	1		2.0	3
.5	1.5		1.5	2.5
.75	1.75		1.25	2.25
.9	1.9		1.1	2.1
.99	1.99		1.01	2.01
.999	1.999		1.001	2.001

It is easy to see that we can get $f(x)$ as close to 2 as we want by picking x sufficiently close to 1. Thus, $\lim_{x\to 1} (x + 1) = 2$.

Problem 2.1.2. What is $\lim_{x\to 3} \dfrac{x(x - 3)}{x - 3}$?

Here we simply cannot enter 3 into the function for x to obtain the limit because doing so would make the denominator equal to 0. Thus, the function is undefined at $x = 3$. Recall, however, that we do not actually care about the value of the function *at* $x = 3$ but only the value the function *approaches* as $x \to 3$. This allows us to cancel the like terms, so we are left with $\lim_{x\to 3} x$. Because this reduced form is linear, it is easy to see that $\lim_{x\to 3} x = 3$. The graph of $f(x) = \dfrac{x(x - 3)}{(x - 3)}$ appears in Figure 2.1. Notice that this graph looks just like the graph of $f(x) = x$ except that it has a "hole" in it at $x = 3$.

Problem 2.1.3. What is $\lim_{x\to -2} \dfrac{x^2 + 4x + 4}{x^2 - 4}$?

This problem is less obvious, but you should see that we cannot just enter -2 into the function because that will result in a 0 in the denominator. We could start by constructing a table of values, but it might be easier to factor the numerator and denominator to see if we can make the function simpler. After factoring the top and bottom, we find

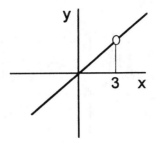

Figure 2.1. Graph for Problem 2.1.2: $f(x) = \dfrac{x(x-3)}{(x-3)}$

$$f(x) = \frac{(x+2)(x+2)}{(x+2)(x-2)} = \frac{(x+2)}{(x-2)}.$$

This function is easier to work with, and we can see that entering -2 into this reduced function yields 0. Thus,

$$\lim_{x \to -2} \frac{x^2 + 4x + 4}{x^2 - 4} = 0.$$

(Confirm this for yourself by constructing a table of values.)

2.2 Infinite Limits and Limits at Infinity

Problem 2.2.1. What is $\lim\limits_{x \to -1} (x+1)^{-4}$?

Begin by rewriting the problem as $\lim\limits_{x \to -1} \dfrac{1}{(x+1)^4}.$

In this form, it is clear that we just cannot enter -1 into the function because it would create a 0 in the denominator. Unlike previous problems, we also cannot factor any terms. If we plot a few points, the graph of the function will appear as in Figure 2.2. It should be clear from this graph that as $x \to -1$, the denominator will get very small. The smaller the denominator, the larger the result. Thus, as $(x+1) \to 0$, $f(x) \to \infty$. More specifically, $\lim\limits_{x \to -1} (x+1)^{-4} = \infty$.

22

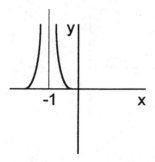

Figure 2.2. Graph for Problem 2.2.1: $\lim_{x \to -1} = (x + 1)^{-4}$

Problem 2.2.2. What is $\lim_{x \to +\infty} (x + 1)^{-4}$?

Here we are interested in determining whether the function has a limit if we allow x to go to positive infinity. As $x \to +\infty$, so does $(x + 1)^4$. Dividing 1 by increasingly larger denominators results in an increasingly small value, that is, $\lim_{x \to +\infty} (x + 1)^{-4} = 0$. We can see this from Figure 2.2.

2.3 Properties of Finite Limits

Recall that if $\lim_{x \to a} f(x)$ and $\lim_{x \to a} g(x)$, both exist and the following rules hold:

1. $\lim_{x \to a} [f(x) \pm g(x)] = \lim_{x \to a} f(x) \pm \lim_{x \to a} g(x)$; addition and subtraction

2. $\lim_{x \to a} [f(x) \times c] = c \times \lim_{x \to a} f(x)$; multiplication by a constant

3. $\lim_{x \to a} [f(x) \times g(x)] = \left(\lim_{x \to a} f(x) \right) \left(\lim_{x \to a} g(x) \right)$; multiplication

4. $\lim_{x \to a} [f(x)/g(x)] = \lim_{x \to a} f(x)/\lim_{x \to a} g(x)$, if $\lim_{x \to a} g(x) \neq 0$; division

5. $\lim_{x \to a} [f(x)]^n = \left(\lim_{x \to a} f(x) \right)^n$, if $\lim_{x \to a} f(x) > 0$; exponent rule

Based on rules 1 and 2, we can add one more rule specifically for linear functions.

6. $\lim\limits_{x \to a} (mx + b) = ma + b$

Problem 2.3.1. Use the rules for limits to evaluate the following.

a. $\lim\limits_{x \to 1} (2x^2 + 1)$: Here we need to find the limit of a polynomial. First use rules 1, 2, and 5 to simplify this limit:

$$\lim_{x \to 1} (2x^2 + 1) = \lim_{x \to 1} 2x^2 + \lim_{x \to 1} 1 = 2 \lim_{x \to 1} x^2 + \lim_{x \to 1} 1$$
$$= 2(\lim_{x \to 1} x^2) + \lim_{x \to 1} 1 = 2(\lim_{x \to 1} x)^2 + \lim_{x \to 1} 1$$

Now use rule 6 to evaluate the two limits of linear functions, $\lim\limits_{x \to 1} x$ and $\lim\limits_{x \to 1} 1$, and finish the calculations:

$$2(\lim_{x \to 1} x)^2 + \lim_{x \to 1} 1 = 2(1)^2 + 1 = 3$$

b. $\lim\limits_{x \to -1} (3x^3 + 4x - 8)$: Use rules 1, 2, and 5 to first break the limit into simpler components, then proceed as before:

$$\lim_{x \to -1} (3x^3 + 4x - 8) = \lim_{x \to -1} 3x^3 + \lim_{x \to -1} 4x - \lim_{x \to -1} 8$$

$$= 3(\lim_{x \to -1} x)^3 + 4(\lim_{x \to -1} x) - \lim_{x \to -1} 8 = 3(-1)^3 + 4(-1) - 8$$

$$= -3 - 4 - 8 = -15$$

c. $\lim\limits_{x \to 0} \dfrac{2x^3 - 8}{x - 1}$: Here we will use rule 4, but to do so we must first be sure that the denominator does not approach 0. Because $\lim\limits_{x \to 0} x - 1 = -1 \neq 0$, we may proceed:

$$\lim_{x \to 0} \frac{2x^3 - 8}{x - 1} = \frac{\lim\limits_{x \to 0} (2x^3 - 8)}{\lim\limits_{x \to 0} (x - 1)} = \frac{2(\lim\limits_{x \to 0} x)^3 - \lim\limits_{x \to 0} 8}{\lim\limits_{x \to 0} x - \lim\limits_{x \to 0} 1} = \frac{0 - 8}{0 - 1} = 8$$

d. $\lim\limits_{x\to 2}\dfrac{x+1}{x-2}$: We cannot apply rule 4 for this problem because $\lim\limits_{x\to 2} x - 2 = 0$. Because the limit of the numerator is not equal to 0 (i.e., $\lim\limits_{x\to 2} x + 1 = 3$), we can conclude that the limit for this quotient does not exist. Figure 2.3 shows why this is true. As we approach $x = 2$ from the left, the function decreases without bound, but as we approach $x = 2$ from the right, the function increases without bound.

e. $\lim\limits_{x\to 2}\dfrac{x^2-4}{x^2-x-2}$: In this problem, as $x \to 2$, both the numerator and denominator go to 0. Begin by factoring and canceling like terms:

$$\lim_{x\to 2}\frac{x^2-4}{x^2-x-2} = \lim_{x\to 2}\frac{(x-2)(x+2)}{(x+1)(x-2)} = \lim_{x\to 2}\frac{(x+2)}{(x+1)}$$

When canceling $x - 2$ from the numerator and denominator, we must remember that $x \neq 2$ (otherwise we would be dividing by 0). Now evaluate the remaining limit:

$$\lim_{x\to 2}\frac{(x+2)}{(x+1)} = \frac{\lim\limits_{x\to 2}(x+2)}{\lim\limits_{x\to 2}(x+1)} = \frac{\lim\limits_{x\to 2}x + \lim\limits_{x\to 2}2}{\lim\limits_{x\to 2}x + \lim\limits_{x\to 2}1} = \frac{2+2}{2+1} = \frac{4}{3}$$

f. $\lim\limits_{x\to 4}\dfrac{\sqrt{x}-2}{x-4}$: Once again, both the numerator and denominator go to 0 as $x \to 4$. To simplify this quotient, we must multiply the top and bottom by $\sqrt{x} + 2$:

$$\lim_{x\to 4}\frac{\sqrt{x}-2}{x-4} = \lim_{x\to 4}\frac{(\sqrt{x}-2)(\sqrt{x}+2)}{(x-4)(\sqrt{x}+2)} = \lim_{x\to 4}\frac{x-4}{(x-4)(\sqrt{x}+2)} = \lim_{x\to 4}\frac{1}{\sqrt{x}+2}$$

This quotient is simpler and we can now take the limit as $x \to 4$:

$$\lim_{x\to 4}\frac{1}{\sqrt{x}+2} = \frac{\lim\limits_{x\to 4}1}{\left(\lim\limits_{x\to 4}x\right)^{1/2}+\lim\limits_{x\to 4}2} = \frac{1}{4^{1/2}+2} = \frac{1}{2+2} = \frac{1}{4}$$

(*Note:* For purposes of this problem I have ignored the negative square root of 2.)

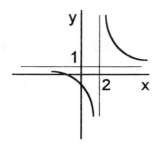

Figure 2.3. Graph for Problem 2.3.1d: $f(x) = \dfrac{x+1}{x-2}$

g. $\displaystyle\lim_{y\to+\infty} \frac{2y^3 + 4}{y^3 + y - 1}$: The use of y as the variable here rather than x is nothing more than a name change, and all the rules apply in the same way. Both the numerator and denominator go to $+\infty$ as $y \to +\infty$, but it does not make sense to think of the limit as some form of infinity divided by infinity. To evaluate this limit, multiply the top and bottom by 1 in the following manner:

$$\lim_{y\to+\infty} \frac{2y^3 + 4}{y^3 + y - 1} = \lim_{y\to+\infty} \frac{(2y^3 + 4)\left(\dfrac{1}{y^3}\right)}{(y^3 + y - 1)\left(\dfrac{1}{y^3}\right)} = \lim_{y\to+\infty} \frac{\dfrac{2y^3}{y^3} + \dfrac{4}{y^3}}{\dfrac{y^3}{y^3} + \dfrac{y}{y^3} - \dfrac{1}{y^3}} =$$

$$\lim_{y\to+\infty} \frac{2 + \dfrac{4}{y^3}}{1 + \dfrac{1}{y^2} - \dfrac{1}{y^3}}$$

Although the resulting limit looks messy, it is much easier to evaluate. The terms with a power of y in the denominator will go to 0 as $y \to +\infty$, so we are left with the following:

$$\lim_{y\to+\infty} \frac{2 + \dfrac{4}{y^3}}{1 + \dfrac{1}{y^2} - \dfrac{1}{y^3}} = \frac{\displaystyle\lim_{y\to+\infty} 2 + \lim_{y\to+\infty} \dfrac{4}{y^3}}{\displaystyle\lim_{y\to+\infty} 1 + \lim_{y\to+\infty} \dfrac{1}{y^2} - \lim_{y\to+\infty} \dfrac{1}{y^3}} = \frac{2 + 0}{1 + 0 - 0} = 2$$

The key to this technique is to divide the numerator and the denominator by the highest power of the variable of interest (in this problem y) found in the denominator.

2.4 Continuity

Recall that for a function to be continuous at $f(a)$, three conditions must be met: (a) $f(a)$ is defined, (b) $\lim_{x \to a} f(x)$ exists, and (c) $f(a) = \lim_{x \to a} f(x)$.

Problem 2.4.1. Which of the functions in Figure 2.4 are continuous at $x = a$?

a. Continuous; all three of the conditions are met.

b. Discontinuous; $\lim_{x \to a} f(x)$ exists, but $f(a)$ is not defined.

c. Discontinuous; $f(a)$ exists, but $\lim_{x \to a} f(x)$ does not.

d. Discontinuous; both $f(a)$ and $\lim_{x \to a} f(x)$ exist, but $\lim_{x \to a} f(x) \neq f(a)$.

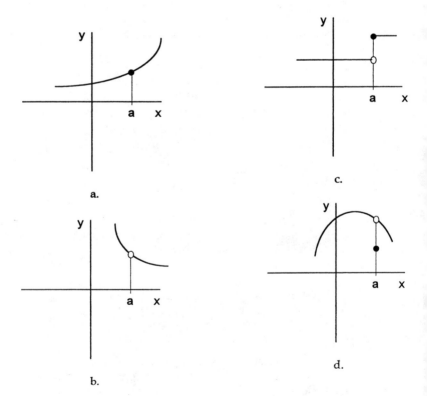

Figure 2.4. Graphs of Functions for Problem 2.4.1

2.5 Homework Problems

1. Does $\lim_{x \to a} f(x)$ exist for the four functions shown in Figure 2.5?

2. Find the following limits if they exist.

    a. $\lim_{x \to 2} \dfrac{1}{x - 2}$ c. $\lim_{x \to \infty} 7(x - 3)^{-1}$

    b. $\lim_{x \to -3} (x + 3)^{-2}$ d. $\lim_{x \to \infty} (2x^2 + 1)^{-1}(3x + 5)^2$

3. Find the following limits if they exist.

    a. $\lim_{x \to 1} 7$ c. $\lim_{x \to 3} (3x^2 - 5x + 2)$

    b. $\lim_{x \to -2} (x + 3)$ d. $\lim_{x \to 3} (x - 1)^2(x + 1)$

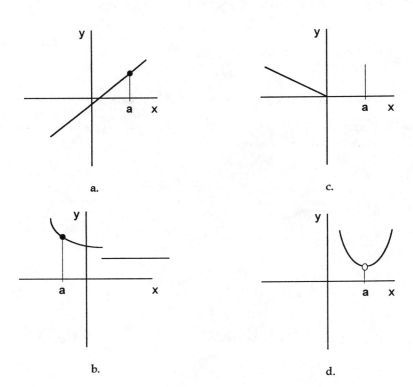

Figure 2.5. Graphs for Homework Problem 2.1

e. $\lim\limits_{x \to -3} \dfrac{3x+9}{x+3}$

h. $\lim\limits_{x \to 4} \dfrac{x^2 - 3x - 4}{x^2 - 5x + 4}$

f. $\lim\limits_{x \to 2} \dfrac{4 - x^2}{x - 2}$

i. $\lim\limits_{x \to -2} \dfrac{x^2 - x - 6}{x^2 + 3x + 2}$

g. $\lim\limits_{x \to 4} \dfrac{x+1}{4-x}$

j. $\lim\limits_{x \to 9} \dfrac{\sqrt{x} - 3}{x - 9}$

4. Where are the following functions continuous?

a. $\dfrac{2x - 4}{3x - 2}$

c. $f(x) = \begin{cases} x+1 & \text{if } -3 \le x < 0 \\ x-1 & \text{if } x \ge 0 \end{cases}$

b. $f(x) = \begin{cases} 3-x & \text{if } x < 2 \\ x-2 & \text{if } x \ge 2 \end{cases}$

3. DIFFERENTIAL CALCULUS

3.1 Tangents to Curves

Problem 3.1.1. Using the formula $\lim\limits_{x_1 \to x_0} \dfrac{y_1 - y_0}{x_1 - x_0}$, determine the slope of the tangent line at $x = 2$ if $f(x) = x^2$. Begin by entering $f(x)$, $f(2)$, x, and 2 into the formula as follows:

$$\lim_{x \to 2} \frac{f(x) - f(2)}{x - 2} = \lim_{x \to 2} \frac{x^2 - 2^2}{x - 2} = \lim_{x \to 2} \frac{x^2 - 4}{x - 2} = \lim_{x \to 2} \frac{(x - 2)(x + 2)}{x - 2}$$

Now simplify and calculate the limit:

$$\lim_{x \to 2} \frac{(x - 2)(x + 2)}{x - 2} = \lim_{x \to 2} (x + 2) = 4$$

Problem 3.1.2. Using the formula $\lim\limits_{\Delta x \to 0} \dfrac{f(x + \Delta x) - f(x)}{\Delta x}$, determine the slope of the tangent line at $x = 3$ if $f(x) = x^2$. Begin by entering $f(x)$ into the formula as follows:

$$\lim_{\Delta x \to 0} \frac{(x + \Delta x)^2 - x^2}{\Delta x} = \lim_{\Delta x \to 0} \frac{(x^2 + 2x\,\Delta x + \Delta x^2) - x^2}{\Delta x} = \lim_{\Delta x \to 0} \frac{2x\,\Delta x + \Delta x^2}{\Delta x}$$

$$= \lim_{\Delta x \to 0} \frac{\Delta x(2x + \Delta x)}{\Delta x} = \lim_{\Delta x \to 0} 2x + \Delta x$$

Now enter $x = 3$. As $\Delta x \to 0$, we see that the slope of the tangent line at $x = 3$ is 6:

$$\lim_{\Delta x \to 0} 2x + \Delta x = \lim_{\Delta x \to 0} 6 + \Delta x = 6$$

Although this is the same function as in 3.1.1, we can see that the slope changes as we move to a different point. These problems demonstrate how to calculate the slope of the tangent line at a particular point, that is, the *derivative* of the function at that point.

3.2 Differentiation Rules

Recall the rules for differentiation:

1. If $f(x) = c$, a constant, then $f'(x) = 0$: The derivative of a constant is 0.

2. If $f(x) = mx + b$, then $f'(x) = m$: The derivative of a linear function is the slope of the line, m.

3. $\frac{d}{dx}[c \times f(x)] = c \times \frac{d}{dx}[f(x)]$: You can pull a constant through the differentiation.

4. If $f(x) = x^n$, then $f'(x) = nx^{n-1}$: This is the power rule.

5. $(f + g)': x = f'(x) + g'(x)$: The derivative of the sum of two functions is equal to the sum of the derivatives of the functions.

6. $(f \times g)': x = [f'(x) \times g(x)] + [f(x) \times g'(x)]$: This is the product rule.

7. $(f/g)': x = \dfrac{[f'(x) \times g(x)] - [f(x) \times g'(x)]}{[g(x)]^2}$, if $g(x) \neq 0$: This is the quotient rule.

8. $(f \circ g)': x = g'(x) \times f'[(g(x)]$: This is the chain rule.

9. $f(x) = e^x \Rightarrow f'(x) = e^x$

   $f(x) = a^x \Rightarrow f'(x) = (\ln a) \times a^x$

10. $f(x) = \ln x \Rightarrow f'(x) = \dfrac{1}{x}$

   $f(x) = \log_a x \Rightarrow f'(x) = \dfrac{1}{x \times \ln a}$

Problem 3.2.1. Find the derivatives of the following functions.

a. $f(x) = -1$: The derivative of a constant function is 0. Thus, $f'(x) = 0$.

b. $f(x) = 3x - 2$: The derivative of a linear function is the slope of the line. Thus, $f'(x) = 3$.

c. $f(x) = 4x + 1$: This is another linear function, so $f'(x) = 4$, but we can apply rules 1, 3, 4, and 5 to reach the same result. Begin by thinking of $f(x)$ as two functions, $g(x) = 4x$ and $h(x) = 1$. From rule 5 we know $(g + h)': x = g'(x) + h'(x)$. Thus,

$$f'(x) = (g + h)': x = g'(x) + h'(x) = \frac{d}{dx}4x + \frac{d}{dx}1$$

From rule 1 we know that $\dfrac{d}{dx}1 = 0$. Using rule 3, we can pull the constant through the differentiation: $\dfrac{d}{dx}4x = 4\dfrac{d}{dx}x$. From the power rule we know $\dfrac{d}{dx}x^n = n\dfrac{d}{dx}x^{n-1}$. Here, $n = 1$, so $\dfrac{d}{dx}x = 1 \times x^0 = 1$. Putting it all together, we have $\dfrac{d}{dx}4x + \dfrac{d}{dx}1 = 4\dfrac{d}{dx}x + 0 = 4 \times 1 = 4$. (Rule 2 is a special case of rules 1, 3, 4, and 5.)

d. $f(x) = 3x^7$: This is an application of rules 3 and 4. We can find the derivative by applying each rule as follows:

$$f'(x) = \frac{d}{dx}3x^7 = 3\frac{d}{dx}x^7 = 3 \times 7x^{7-1} = 21x^6$$

As you become more comfortable with the rules, you can begin to combine steps.

e. $g(x) = 5x^2 + 2$: Using g rather than f to represent the function makes no difference. Applying rules 1, 3, and 4, we find that $g'(x) = 10x$.

f. $h(t) = 6t^3 - 4$: Again, the use of h rather than f to represent the function makes no difference. The use of the variable t rather than x also does not matter, *provided* that when we take the derivative we do so with respect to t. Applying rules 1, 3, and 4 to the rest, we find $h'(t) = 18t^2$.

g. $f(x) = x^3 + \dfrac{1}{x^2}$: We will again use the power rule, but it might be easier to see by rewriting the function as $f(x) = x^3 + x^{-2}$. Now apply the power rule to each term:

$$f'(x) = 3x^{3-1} + (-2)x^{-2-1} = 3x^2 - 2x^{-3} = 3x^2 - \frac{2}{x^3}$$

As you can see, the power rule also applies to negative exponents. Remember, however, that subtracting 1 from a negative number results in a negative number of *larger* magnitude (e.g., $-2 - 1 = -3$).

h. $h(x) = x^{2/3} - 3x^{-1}$: The power rule also applies to fractional exponents. Apply the power rule to each term as follows:

$$h'(x) = \frac{2}{3}x^{2/3-1} - 3(-1)x^{-1-1} = \frac{2}{3}x^{-1/3} + 3x^{-2}$$

Problem 3.2.2. Use the product rule to find the derivatives of the following functions.

a. $f(x) = (x + 1)(x - 1)$: We could just multiply the two factors and take the derivative of the result, but let us apply the product rule. Think of $g(x) = x + 1$ and $h(x) = x - 1$ and apply the product rule as follows:

$$f'(x) = (g \times h)': x = [g'(x) \times h(x)] + [g(x) \times h'(x)]$$
$$= 1 \times (x - 1) + (x + 1) \times 1 = (x - 1) + (x + 1) = 2x$$

You should see that you get the same derivative as if you had multiplied the two factors first.

b. $f(t) = (3t^2 - 5t - 7)(t^2 + 2t - 3)$: Here, think of $g(t) = (3t^2 - 5t - 7)$ and $h(t) = (t^2 + 2t - 3)$ and apply the product rule as follows:

$$f'(t) = g'(t)h(t) + g(t)h'(t) = (6t - 5)(t^2 + 2t - 3) + (3t^2 - 5t - 7)(2t + 2)$$
$$= 6t^3 + 12t^2 - 18t - 5t^2 - 10t + 15 + 6t^3 + 6t^2 - 10t^2 - 10t - 14t - 14$$
$$= 12t^3 + 3t^2 - 52t + 1$$

c. $g(t) = (6t^{4/3} - 7t)(3t + t^{-3/5})$: The presence of negative and fractional exponents in the terms does not affect the application of the product rule. Because the function is labeled g, think of the first term as $f(t)$ and the second term as $h(t)$ and apply the product rules as follows:

$$g'(t) = f'(t)h(t) + f(t)h'(t)$$
$$= [6(\tfrac{4}{3})t^{\frac{4}{3}-1} - 7](3t + t^{-\frac{3}{5}}) + (6t^{\frac{4}{3}} - 7t)[3 + (-\tfrac{3}{5})t^{-\frac{3}{5}-1}]$$
$$= (8t^{\frac{1}{3}} - 7)(3t + t^{-\frac{3}{5}}) + (6t^{\frac{4}{3}} - 7t)(3 - \tfrac{3}{5}t^{-\frac{8}{5}})$$

Because this is such a messy problem, it is unlikely that you would be asked to do the final multiplications. Even so, for practice you might want to do them to find that $g'(t) = 42t^{\frac{4}{3}} - 42t + \dfrac{22}{5}t^{\frac{4}{15}} - \dfrac{14}{5}t^{-\frac{3}{5}}$.

Problem 3.2.3. Use the quotient rule to find the derivatives of the following functions.

a. $f(x) = \dfrac{1}{x^2}$: Alhough it would be easier to find the derivative of this function using the power rule, apply the quotient rule to see that we get the same answer. Think of $g(x) = 1$ and $h(x) = x^2$ (not x^{-2}) and apply the quotient rule as follows:

$$f'(x) = (g/h)': x = \dfrac{[g'(x)h(x)] - [g(x)h'(x)]}{[h(x)]^2}$$

$$= \dfrac{(0 \times x^2) - (1 \times 2x)}{(x^2)^2} = \dfrac{0 - 2x}{x^4} = \dfrac{-2}{x^3} = -2x^{-3}$$

Notice that this is the same derivative as in the second term of Problem 3.2.1g.

b. $f(t) = \dfrac{t+1}{t-2}$: Think of $g(t) = t + 1$ and $h(t) = t - 2$ and proceed as follows:

$$f'(t) = \dfrac{[g'(t)h(t)] - [g(t)h'(t)]}{[h(t)]^2} = \dfrac{[1 \times (t - 2)] - [(t + 1) \times 1]}{(t - 2)^2}$$

$$= \dfrac{(t - 2) - (t + 1)}{(t - 2)^2} = \dfrac{-3}{(t - 2)^2}$$

c. $h(x) = \dfrac{1-x^3}{x}$: Think of $f(x) = 1 - x^3$ and $g(x) = x$, then:

$$h'(x) = \frac{[f'(x)g(x)] - [f(x)g'(x)]}{[g(x)]^2} = \frac{(-3x^2 \times x) - [(1 - x^3) \times 1]}{x^2}$$

$$= \frac{-3x^3 - (1 - x^3)}{x^2} = \frac{-3x^3 - 1 + x^3}{x^2} = \frac{-2x^3 - 1}{x^2}$$

Although I asked you to use the quotient rule on this problem, it would have been easier to have simplified the original problem to $h(x) = x^{-1} - x^2$ and taken the derivative of each term.

Problem 3.2.4. Use the chain rule to find the derivatives of the following functions.

a. $f(x) = (x^2 - 2)^3$: Think of $f(x)$ as being the composition of two functions, $g(y) = y^3$ and $h(x) = x^2 - 2$. Recall that the new variable y is just to help in distinguishing the two functions. Now apply the chain rule as follows:

$$f(x) = g_3 \circ h: x$$
$$g(y) = y^3$$
$$h(x) = x^2 - 2$$
$$f'(x) = (g \circ h)': x = h'(x)g'[h(x)]$$
$$h'(x) = 2x$$
$$g'(y) = 3y^2$$
$$f'(x) = 2x \times 3(x^2 - 2)^2 = 6x(x^2 - 2)^2$$

b. $f(x) = (5x^3 - x^2)^{-3}$: Allow $g(y) = y^{-3}$ and $h(x) = 5x^3 - x^2$ and we find the following:

$$f(x) = g \circ h: x$$
$$g(y) = y^{-3}$$
$$h(x) = 5x^3 - x^2$$
$$f'(x) = (g \circ h)': x = h'(x)g'[h(x)]$$
$$g'(y) = -3y^{-4}$$
$$h'(x) = 15x^2 - 2x$$
$$f'(x) = (15x^2 - 2x) \times -3(5x^3 - x^2)^{-4} = \frac{-3(15x^2 - 2x)}{(5x^3 - x^2)^4}$$

c. $f(x) = (2x^4 + 6x^{-3})^{-\frac{1}{2}}$: Allow $g(y) = y^{-\frac{1}{2}}$ and $h(x) = 2x^4 + 6x^{-3}$ and we find the following:

$$f(x) = g \circ h: x$$
$$g(y) = y^{-\frac{1}{2}}$$

$$h(x) = 2x^4 + 6x^{-3}$$
$$f'(x) = (g \circ h)': x = h'(x)g'[h(x)]$$
$$g'(y) = -\tfrac{1}{2}y^{-\tfrac{3}{2}}$$
$$h'(x) = 8x^3 - 18x^{-4}$$
$$f'(x) = (8x^3 - 18x^{-4}) \times -\tfrac{1}{2}(2x^4 + 6x^{-3})^{-\tfrac{3}{2}} = \frac{-(8x^3 - 18x^{-4})}{2(2x^4 + 6x^{-3})^{\tfrac{3}{2}}}$$

Problem 3.2.5. Use the differentiation rules for functions with e to find the derivatives of the following functions.

a. $f(x) = 4e^x$: Apply rule 3 by pulling the constant 4 through the differentiation. By rule 9, the derivative of e^x is just e^x, so $f'(x) = 4e^x$.

b. $f(t) = 2t^3e^t$: Think of $g(t) = 2t^3$ and $h(t) = e^t$ and apply the product rule:

$$f'(t) = g'(t)h(t) + g(t)h'(t) = 6t^2 \times e^t + 2t^3 \times e^t$$
$$= 6t^2e^t + 2t^3e^t = 2t^2e^t(t + 3)$$

c. $f(x) = e^{4x^3}$: This function requires the chain rule. Think of $g(y) = e^y$ and $h(x) = 4x^3$ and apply the chain rule as follows:

$$f'(x) = (g \circ h)': x = h'(x)g'[h(x)]$$
$$g'(y) = e^y$$
$$h'(x) = 12x^2$$
$$f'(x) = 12x^2 \times e^{4x^3} = 12x^2e^{4x^3}$$

d. $f(x) = 7x^{-2}e^{3x^2}$: This problem requires the use of both the product rule and the chain rule. To apply the product rule to $7x^{-2}$ and e^{3x^2}, we will need to know the derivative of e^{3x^2}, so begin by using the chain rule as follows:

$$r(x) = e^{3x^2}$$
$$s(y) = e^y$$
$$t(x) = 3x^2$$
$$r'(x) = t'(x)s'[t(x)]$$
$$s'(y) = e^y$$
$$t'(x) = 6x$$
$$r'(x) = 6x \times e^{3x^2} = 6xe^{3x^2}$$

I used the letters r, s, and t to represent the functions above only to distinguish them from the ones I will use in the next step. Now that we know the derivative of e^{3x^2}, we can think of $g(x) = 7x^{-2}$ and $h(x) = e^{3x^2}$ and apply the product rule:

$$f'(x) = g'(x)h(x) + g(x)h'(x)$$
$$= -14x^{-3} \times e^{3x^2} + 7x^{-2} \times 6xe^{3x^2}$$
$$= -14x^{-3}e^{3x^2} + 42x^{-1}e^{3x^2}$$

e. $f(x) = \dfrac{8e^{x^{-5}}}{(5x^2 + 4)^3}$: For this problem we will need to use the quotient rule as well as the chain rule on both the numerator and the denominator. As before, begin by using the chain rule:

Numerator:
$r(x) = 8e^{x^{-5}}$
$s(y) = 8e^y$
$t(x) = x^{-5}$
$r'(x) = t'(x)s'[t(x)]$
$s'(y) = 8e^y$
$t'(x) = -5x^{-6}$
$r'(x) = -5x^{-6} \times 8e^{x^{-5}}$
$\quad\ = -40x^{-6}e^{x^{-5}}$

Denominator:
$r(x) = (5x^2 + 4)^3$
$s(y) = y^3$
$t(x) = 5x^2 + 4$
$r'(x) = t'(x)s'[t(x)]$
$s'(y) = 3y^2$
$t'(x) = 10x$
$r'(x) = 10x \times 3(5x^2 + 4)^2$
$\quad\ = 30x(5x^2 + 4)^2$

Notice that I used the letters r, s, and t for both sets. Again, the letters do not matter as long as you can follow the procedures. Now apply the quotient rule to the original function, $f(x)$, and let $g(x) = 8e^{x^{-5}}$ and $h(x) = (5x^2 + 4)^3$:

$$f'(x) = \frac{g'(x)h(x) - g(x)h'(x)}{[h(x)]^2}$$

$$= \frac{-40x^{-6}e^{x^{-5}} \times (5x^2 + 4)^3 - 8e^{x^{-5}} \times 30x(5x^2 + 4)^2}{[(5x^2 + 4)^3]^2}$$

$$= \frac{-40x^{-6}e^{x^{-5}}(5x^2 + 4)^3 - 240xe^{x^{-5}}(5x^2 + 4)^2}{(5x^2 + 4)^6}$$

$$= \frac{-40x^{-6}e^{x^{-5}}(5x^2 + 4) - 240xe^{x^{-5}}}{(5x^2 + 4)^4}$$

$$= \frac{-40e^{x^{-5}}(5x^{-4} + 4x^{-6} + 6x)}{(5x^2 + 4)^4}$$

Problem 3.2.6. Use the differentiation rules for logarithmic functions to find the derivatives of the following functions.

a. $g(x) = \ln x^2$: Here we will use our ability to manipulate logarithmic functions along with differentiation rule 10 to find the derivative:

$$g(x) = \ln x^2 = 2 \ln x$$
$$g'(x) = 2 \times \frac{1}{x} = \frac{2}{x}$$

b. $f(t) = 7t^{-3}(\ln t)$: This problem requires the use of the product rule. Think of $g(t) = 7t^{-3}$ and $h(t) = \ln t$ and proceed as follows:

$$f'(t) = g'(t)h(t) + g(t)h'(t)$$
$$= -21t^{-4} \times (\ln t) + 7t^{-3} \times \frac{1}{t} = -21t^{-4}(\ln t) + 7t^{-4}$$

c. $f(s) = \ln(6s^2 + 4s)$: This problem requires the use of the chain rule. Think of $g(t) = \ln t$ (just to use a different variable letter) and $h(s) = 6s^2 + 4s$ and apply the chain rule as follows:

$$f'(s) = h'(s)g'[h(s)]$$
$$g'(t) = \frac{1}{t}$$
$$h'(s) = 12s + 4$$
$$f'(s) = (12s + 4) \times \frac{1}{6s^2 + 4s} = \frac{12s + 4}{6s^2 + 4s} = \frac{6s + 2}{3s^2 + 2s}$$

Problem 3.2.7. Find the second derivatives of the following functions.

a. $f(x) = 7x + 2$: Find the first derivative and then take the derivative of the first derivative to obtain the second derivative of $f(x)$:

$$f(x) = 7x + 2$$
$$f'(x) = 7$$
$$f''(x) = 0$$

Recall that the second derivative of a function $f(x)$ can be represented as $f''(x), f^2(x)$, or f_{xx} (among other ways).

b. $g(x) = 3x^{-2/3}$: Use the power rule as follows:

$$g'(x) = 3 \times \frac{-2}{3}x^{-\frac{2}{3}-1} = -2x^{-\frac{5}{3}}$$

$$g''(x) = (-2) \times \frac{-5}{3}x^{-\frac{5}{3}-1} = \frac{10}{3}x^{-\frac{8}{3}}$$

c. $f(x) = (2x^4 + 9x)^3$: This problem will require two applications of the chain rule. Think of $g(y) = y^3$ and $h(x) = 2x^4 + 9x$ to find the first derivative as follows:

$$f'(x) = h'(x)g'[h(x)]$$
$$g'(y) = 3y^2$$
$$h'(x) = 8x^3 + 9$$
$$f'(x) = (8x^3 + 9) \times 3(2x^4 + 9x)^2 = 3(8x^3 + 9)(2x^4 + 9x)^2$$

To find the second derivative, we will need to use the product and chain rules. Start by finding the derivative of $(2x^4 + 9x)^2$. Think of $g(y) = y^2$ and $h(x) = 2x^4 + 9x$ and proceed as follows:

$$\frac{d}{dx}(2x^4 + 9x)^2 = h'(x)g'[h(x)]$$

$$g'(y) = 2y$$
$$h'(x) = 8x^3 + 9$$
$$\frac{d}{dx}(2x^4 + 9x)^2 = (8x^3 + 9) \times 2(2x^4 + 9x)$$
$$= 2(8x^3 + 9)(2x^4 + 9x)$$

Notice that I used the function letters g and h again without indicating that I was working on the second derivative. There was no reason to do so because I was treating $(2x^4 + 9x)^2$ as any other function. I could have used different letters for the functions, but given the multiple stages for this problem I would have so many letters that the problem would become even more confusing. Now that we know the derivative of $(2x^4 + 9x)^2$, we can use the product rule to finish the problem. Think of $g(x) = 8x^3 + 9$ and $h(x) = (2x^4 + 9x)^2$ and proceed as follows:

$$\frac{d}{dx}(8x^3 + 9)(2x^4 + 9x)^2 = g'(x)h(x) + g(x)h'(x)$$
$$= 24x^2 \times (2x^4 + 9x)^2 + (8x^3 + 9) \times [2(8x^3 + 9)(2x^4 + 9x)]$$
$$= 24x^2(2x^4 + 9x)^2 + 2(8x^3 + 9)^2(2x^4 + 9x)$$

I left the constant 3 out of the above calculations to simplify things a bit. To get the final answer, we need to multiply the last line above by 3:

$$f''(x) = 3[24x^2(2x^4 + 9x)^2 + 2(8x^3 + 9)^2(2x^4 + 9x)]$$
$$= 72x^2(2x^4 + 9x)^2 + 6(8x^3 + 9)^2(2x^4 + 9x)$$

d. $f(t) = e^{t^2}$: Use the chain rule to find the first derivative. As usual, think of $g(u) = e^u$ and $h(t) = t^2$. Then we find the first derivative to be the following:

$$f'(t) = h'(t)g'[h(t)]$$
$$g'(u) = e^u$$
$$h'(t) = 2t$$
$$f'(t) = 2t(e^{t^2}) = 2t \; e^{t^2}$$

Apply the product rule to find the second derivative. Think of $g(t) = 2t$ and $h(t) = e^{t^2}$. We already know $h'(t)$, so we can proceed directly to the product rule:

$$\frac{d}{dt} 2t \; (e^{t^2}) = g' \; (t)h(t) + g(t)h'(t)$$
$$= 2 \times e^{t^2} + 2t \times 2t \; (e^{t^2})$$
$$= 2e^{t^2} + 4t^2 \; (e^{t^2})$$
$$= 2e^{t^2} \; (1 + 2t^2)$$

Problem 3.2.8. Find the first derivatives of the following functions and calculate the slope of the tangent line for the indicated value.

a. $f(x) = 4x^2 - 3x + 5$; $x = 2$: Begin by finding the derivative of $f(x)$:

$$f'(x) = 8x - 3$$

To find the slope of the line tangent to the function at $x = 2$, enter that value into the derivative function (i.e., $f'(x)$):

$$f'(2) = (8 \times 2) - 3 = 16 - 3 = 13$$

b. $g(x) = 2x^{-1}$; $x = -3$: First use the power rule to find the derivative of the function, then evaluate the derivative at $x = -3$:

$$g'(x) = 2(-1)x^{-1-1} = -2x^{-2}$$
$$g'(-3) = -2(-3)^{-2} = \frac{-2}{(-3)^2} = \frac{-2}{9}$$

c. $f(t) = e^{2t}$; $t = 0$: Again, apply the chain rule to find the derivative, then enter the value into it:

$$f'(t) = 2e^{2t}$$
$$f'(0) = 2e^{2 \times 0} = 2e^0 = 2$$

3.3 Extrema of Functions

Recall that the second derivative test examines the second derivative of the function evaluated at the critical point(s). If $f''(x_0) < 0$, then the critical point x_0 is a relative maximum. If $f''(x_0) > 0$, then x_0 is a relative minimum. If $f''(x_0) = 0$, then the test is inconclusive.

Problem 3.3.1. Find the maxima and minima, if any, of the following functions.

a. $f(x) = x^2 - 4$: To find the extrema of a function we find the first derivative of the function, set it equal to 0, and solve for the variable of interest:

$$f(x) = x^2 - 4$$
$$f'(x) = 2x = 0$$
$$\Rightarrow x = 0$$

Entering this value into the function yields the y-value for the critical point, $f(0) = 0^2 - 4 = -4$. Thus, $(0, -4)$ is a critical point, but we still need to determine whether it is a maximum, minimum, or a point of inflection. It should be relatively clear that $f(x) \to \infty$ as $x \to \pm\infty$ (i.e., as x grows without bound in either the positive or negative direction). Thus, $(0, -4)$ must be a minimum, but let us check using the second derivative test. Begin by calculating $f''(x)$ and then evaluate it at the critical point:

$$f'(x) = 2x$$
$$f''(x) = 2$$

There was actually nothing to evaluate for this function because the second derivative was a constant. Because $f''(x)$ is greater than 0, from the second part of the test we know the critical point must be a minimum.

b. $f(x) = x^3 - 12x$: Find the first derivative, set it equal to 0, and solve for x:

$$f(x) = x^3 - 12x$$
$$f'(x) = 3x^2 - 12 = 0$$
$$\Rightarrow 3(x^2 - 4) = 3(x + 2)(x - 2) = 0$$
$$\Rightarrow x = -2, 2$$

Entering these values into $f(x)$, we find the critical points to be $(-2, 16)$ and $(2, -16)$. Now apply the second derivative test as follows:

$$f'(x) = 3x^2 - 12$$
$$f''(x) = 6x$$
$$f''(-2) = -12 < 0$$
$$f''(2) = 12 > 0$$

Thus, $(-2, 16)$ is a local maximum and $(2, -16)$ is a local minimum. (We know they are not global extrema because the sign on the highest power of x indicates that $f(x) \rightarrow +\infty$ as $x \rightarrow +\infty$ and $f(x) \rightarrow -\infty$ as $x \rightarrow -\infty$.)

c. $f(x) = x^3$: Find the first derivative, set it equal to 0, and solve for x:

$$f(x) = x^3$$
$$f'(x) = 3x^2 = 0$$
$$\Rightarrow x^2 = x \times x = 0$$
$$\Rightarrow x = 0$$

Thus, the critical point is $(0, 0)$. Now apply the second derivative test:

$$f'(x) = 3x^2 = 0$$
$$f''(x) = 6x$$
$$f''(0) = 0$$

Because the second derivative of the function at the critical point is equal to 0, the test gives us no information about the function at that point, and we must turn to the first derivative test. In the first derivative test, we examine the sign of $f'(x)$ in the neighborhood around the critical point. Let us check the points $(-1, -1)$ and $(1, 1)$:

$$f'(x) = 3x^2$$
$$f'(-1) = 3 > 0$$
$$f'(1) = 3 > 0$$

Because the sign of $f'(x)$ does not change as we move from one side of $(0, 0)$ to the other, $(0, 0)$ is a point of inflection. (See Figure 1.1b for the graph of this function.)

Problem 3.2.2. Find any extrema for the following functions on the specified interval:

a. $f(x) = 3x - 4; -3 \leq x \leq 3$: Linear functions do not have extrema when x is unbounded. Here, however, x can take on only the specified values, so we must evaluate the function at the endpoints. Because this is a linear function with a positive slope, it is easy to see that $(-3, -13)$ is a minimum and $(3, 5)$ is a maximum over the specified interval.

b. $f(x) = \frac{1}{3}x^3 - x^2 - 3x; 0 \leq x \leq 4$: Find the first derivative and continue as usual:

$$f(x) = \frac{1}{3}x^3 - x^2 - 3x$$
$$f'(x) = x^2 - 2x - 3 = (x + 1)(x - 3) = 0$$
$$\Rightarrow x = -1, 3$$

The value $x = -1$ lies outside the specified interval, so we only need to check the endpoints and the one critical point $(3, -9)$. The two endpoints are $(0, 0)$ and $\left(4, \frac{-20}{3}\right)$. The endpoint $(0, 0)$ is a *global* maximum on the interval (i.e., no other values of $f(x)$ on the interval are greater than 0), the endpoint $\left(4, \frac{-20}{3}\right)$ is a *relative* maximum on the interval (i.e., no other values of $f(x)$ in the neighborhood of $x = 4$ are greater than $\frac{-20}{3}$), and the critical point $(3, -9)$ is a global minimum on the interval.

Problem 3.3.3. An educational specialist estimates that the average scores on a college's entrance test x years from now will be $f(x) = 732 - 5x$, where 732 is the current average score (out of a possible 1,000). What is the rate at which the average score is changing? What will the average score be 8 years from now?

Determining the estimated average score 8 years from now is simply a matter of entering that value into the equation. Specifically, $f(x) = 732 - 5(8) = 732 - 40 = 692$. Because the change in the average scores follows a linear function, we know the rate of change in the scores is $f'(x) = -5$, a constant.

Problem 3.3.4. An efficiency expert determines that the daily output of a manufacturing plant is related to the number of workers in the plant by the function $Q(x) = 400x - x^2$, where $Q(x)$ is the quantity of items produced and x is the number of workers. There are currently 150 workers. Will the efficiency expert recommend a change in the number of workers (and if so how many)? How much will the addition of 1 worker change the daily output? What is the significance of the value of $Q''(x)$?

Begin by calculating the first and second derivatives of $Q(x)$:

$$Q(x) = 400x - x^2$$
$$Q'(x) = 400 - 2x = 0$$
$$\Rightarrow x = 200$$
$$Q''(x) = -2$$

By setting the first derivative of the function equal to 0 and solving for x, we see that the function has an extremum at $x = 200$. From the second derivative test, we know that (200, 40,000) must be a maximum. Thus, daily output can be increased by hiring an additional 50 workers. Even without knowing the specific number of workers needed to maximize output, the efficiency expert would know that more workers would increase output. By entering 150 into $Q'(x)$, we find the slope of the line tangent to the output function for 150 workers: $Q'(x) = 400 - 2(150) = 100$. Because this number is positive, we know that output would increase if another worker were hired. In fact, one additional worker at this level would increase daily output by 99 units. The addition of a second worker would increase output only by 97 units. This decrease in additional output as workers are added reflects the negative value of the second derivative, $Q''(x)$. In other words, although output will continue to increase with the addition of each worker (up to 200), the *marginal* or per worker increase is decreasing.

Problem 3.3.5. You are to construct a box from a piece of cardboard by cutting a square from each corner and turning up the sides. If the cardboard measures 24 inches by 9 inches, what is the largest volume possible for the box?

The key to word problems such as this is to properly set up the equation. We are looking for the volume of a box, so we will need values for the length, width, and height of the box. We do not know the size of the square to be cut from each corner, which will be the height of the box, so call it x. From Figure 3.1 we can see that the base of the box will be the large, rectangular, darkly shaded area. The length of this area will be $24 - 2x$, the original length of the cardboard minus twice the dimension of the square to be removed. Similarly, the width of the base will be $9 - 2x$. Thus, the volume of the box will be determined by the function:

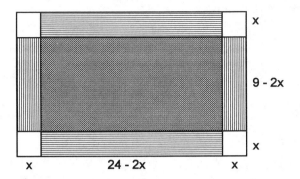

Figure 3.1. Figure for Problem 3.3.3

$$f(x) = x(24 - 2x)(9 - 2x)$$
$$= x(216 - 18x - 48x + 4x^2)$$
$$= 4x^3 - 66x^2 + 216x$$

To maximize the volume, find the derivative of the function, set it equal to 0, and solve for x:

$$f(x) = 4x^3 - 66x^2 + 216x$$
$$f'(x) = 12x^2 - 132x + 216 = 0$$
$$\Rightarrow x^2 - 11x + 18 = 0 \Rightarrow (x - 2)(x - 9) = 0$$
$$\Rightarrow x = 2, 9$$

The two critical points are $(2, 200)$ and $(9, -486)$. The y-value represents volume, so we can ignore the negative value. Thus, cutting two-inch squares from each corner yields the box with the largest volume. Verify this using the second derivative test:

$$f'(x) = 12x^2 - 132x + 216 = 0$$
$$f''(x) = 24x - 132$$
$$f''(2) = 24(2) - 132 = -84 < 0$$
$$f''(9) = 24(9) - 132 = 84 > 0$$

Because the second derivative of $f(x)$ evaluated at $x = 2$ is negative, the point $(2, 200)$ is a maximum.

3.4 Homework Problems

1. Use the formula $\lim\limits_{x_1 \to x_0} \dfrac{y_1 - y_0}{x_1 - x_0}$ to determine the slope of the tangent line at $x = 2$ if $f(x) = 3x^2 - x$.

2. Find the derivatives of the following functions.

   a. $f(x) = -5$

   b. $g(x) = -7x + 6$

   c. $f(t) = 4t^3 - 7t + 2$

   d. $f(x) = x^4 - \dfrac{1}{x^2}$

   e. $g(x) = 3x^{-1} + x^{-5}$

   f. $h(x) = 9x^{7/9} - 2x^{-3/4} + 5x^{-3.2}$

3. Find the derivatives of the following functions.

   a. $g(x) = (2x - 5)(x^2 - 9)$

   b. $f(t) = (t^5 + t)(2t^2 + 6t^{-3})$

   c. $h(x) = (x^2 + 8x - 4)(6x^2 - x - 9)$

   d. $h(t) = (4t^{1/2} + 3t^{1/3})(2t^{-1/2} - 6t^{2/3})$

4. Find the derivatives of the following functions.

   a. $g(t) = \dfrac{3t}{t+7}$

   b. $f(t) = \dfrac{t^2 - 4}{t^4}$

   c. $h(x) = \dfrac{x^2 + 7x - 4}{x^3 - 1}$

   d. $g(x) = 4x + \dfrac{x^3}{1+x^2}$

5. Find the derivatives of the following functions.

   a. $h(x) = (x^2 - 9)^3$

   b. $h(t) = (3t^4 + t - 7)^2$

   c. $g(t) = (t^6 + 8t)^{1/2}$

   d. $f(x) = (6x + x^2 - 2x^3)^{-3/4}$

6. Find the derivatives of the following functions.

   a. $f(x) = e^{2x}$

   b. $g(x) = e^{x^3}$

   c. $h(x) = e^{2x^4 - x^2}$

   d. $f(t) = 6t^2 e^{5t^3}$

   e. $g(t) = (7t^2 - 2)e^{t^2}$

   f. $h(t) = \dfrac{e^t}{(1 + 5t)^2}$

7. Find the derivatives of the following functions.

   a. $h(x) = \ln x^3$

   b. $g(x) = (4x^5 - 3) \ln x$

   c. $f(x) = (\ln x)^3$

   d. $f(t) = \dfrac{\ln x^2}{(1 - x^4)}$

8. Find the first and second derivatives of the following functions.

   a. $f(x) = 3x^3 - 1$

   b. $g(x) = (x + 1)^{-1}$

   c. $g(t) = (2 + t)e^{2t}$

   d. $h(t) = e^{t^2 - 1}$

9. Find the first derivatives of the following functions and calculate the slope of the tangent line at the indicated value.

   a. $h(x) = 4x^2 - 3; x = 2$

   b. $h(t) = (1 - 2t^2)^2; t = 1$

   c. $g(x) = \dfrac{5 + x + x^2}{1 - x^3}; x = -1$

   d. $g(t) = e^{t^3}; t = 1$

10. Find the extrema, if any, of the following functions.

    a. $f(x) = 3 - x^2$

    b. $f(t) = t^3 + t^2 - t$

    c. $g(t) = t^5$

    d. $h(t) = e^{t^2}$

11. Find any extrema for $f(x) = x^3 - 3x$ on the interval $0 \le x \le 2$.

12. The circulation manager of a national magazine estimates that x years from now the number of subscribers will be $S(x) = 50x^2 + 100x + 10,000$, where x is the number of years: (a) What is the rate of change for the number of subscriptions? (b) What will the rate of change be for the fourth year? How much will it actually change during the fifth year? (c) Why is this function overly optimistic?

13. A sociologist wants to mail a survey to 3,000 people. Experience has shown that fewer people respond to longer surveys. In particular, the expected response rate will be $f(p) = 3000 - 40p^2$, where p is the number of pages in the survey instrument. Ten questions are contained on each page of the survey instrument. How many pages should the survey instrument contain to maximize the number of questions for which the sociologist obtains a response?

14. A piece of posterboard for a sign contains 12 square feet. The top and bottom margins are 4 inches each and the side margins are 3 inches. What are the dimensions of the posterboard that maximize the printed area? *Hints:* (a) It will be easier to work in inches. (b) If you know the area and call one dimension x, what is the other dimension in terms of x?

4. MULTIVARIATE FUNCTIONS
AND PARTIAL DERIVATIVES

4.1 Partial Derivatives

Problem 4.1.1. Find the first-order partial derivatives (partials) of $f(x,y) = 2x^3 + y^2$.

Using the notation for partials, we are looking for $\dfrac{\partial f(x, y)}{\partial x}$ and $\dfrac{\partial f(x, y)}{\partial y}$. Recall that when calculating the partial derivative with respect to one variable, we treat all appearances of the other variable(s) as constants. In all other respects, the rules of differentiation apply as usual. Here, in looking for $\dfrac{\partial f(x, y)}{\partial x}$, we treat all appearances of the variable y as a constant. Thus,

$$\frac{\partial f(x, y)}{\partial x} = \frac{\partial}{\partial x}(2x^3 + y^2) = \frac{\partial}{\partial x}2x^3 + \frac{\partial}{\partial x}y^2 = 3 \times 2x^{3-1} + 0 = 6x^2$$

In looking for $\dfrac{\partial f(x, y)}{\partial y}$, y becomes the variable of interest and all appearances of the variable x are treated as constants. Thus,

$$\frac{\partial f(x, y)}{\partial y} = \frac{\partial}{\partial y}(2x^3 + y^2) = \frac{\partial}{\partial y}2x^3 + \frac{\partial}{\partial y}y^2 = 0 + 2y^{2-1} = 2y$$

Problem 4.1.2. Find the first-order partials of $f(x, y) = 3x^4y^6$.

Again, when taking the partial with respect to one variable, treat the other as a constant. Thus,

$$\frac{\partial f(x, y)}{\partial x} = \frac{\partial}{\partial x}3x^4y^6 = y^6 \times \frac{\partial}{\partial x}3x^4 = y^6(4 \times 3x^{4-1}) = y^6(12x^3) = 12x^3y^6$$

$$\frac{\partial f(x, y)}{\partial y} = \frac{\partial}{\partial y}3x^4y^6 = 3x^4 \times \frac{\partial}{\partial y}y^6 = 3x^4(6y^5) = 18x^4y^5$$

Problem 4.1.3. Find the first-order partials of the following functions.

a. $f(x, y) = 2x^7 + 9xy$:

$$\frac{\partial f(x, y)}{\partial x} = \frac{\partial}{\partial x}(2x^7 + 9xy) = \frac{\partial}{\partial x}2x^7 + \frac{\partial}{\partial x}9xy$$

$$= 14x^6 + 9y\frac{\partial}{\partial x}x = 14x^6 + 9y \times 1 = 14x^6 + 9y$$

$$\frac{\partial f(x, y)}{\partial y} = \frac{\partial}{\partial y}(2x^7 + 9xy) = \frac{\partial}{\partial y}2x^7 + \frac{\partial}{\partial y}9xy$$

$$= 0 + 9x \times \frac{\partial}{\partial y}y = 0 + 9x \times 1 = 9x$$

b. $f(s, t) = s^3t^{-2} + 8st - 4s^{-3}t^4$:

$$\frac{\partial f(s, t)}{\partial s} = \frac{\partial}{\partial s}(s^3t^{-2} + 8st - 4s^{-3}t^4) = \frac{\partial}{\partial s}s^3t^{-2} + \frac{\partial}{\partial s}8st - \frac{\partial}{\partial s}4s^{-3}t^4$$

$$= t^{-2} \times \frac{\partial}{\partial s}s^3 + 8t \times \frac{\partial}{\partial s}s - 4t^4 \times \frac{\partial}{\partial s}s^{-3} = t^{-2}(3s^2) + 8t \times 1 - 4t^4(-3s^{-4})$$

$$= 3s^2t^{-2} + 8t + 12s^{-4}t^4$$

$$\frac{\partial f(s, t)}{\partial t} = \frac{\partial}{\partial t}(s^3t^{-2} + 8st - 4s^{-3}t^4) = \frac{\partial}{\partial t}s^3t^{-2} + \frac{\partial}{\partial t}8st - \frac{\partial}{\partial t}4s^{-3}t^4$$

$$= s^3 \times \frac{\partial}{\partial t}t^{-2} + 8s \times \frac{\partial}{\partial t}t - 4s^{-3} \times \frac{\partial}{\partial t}t^4 = s^3(-2t^{-3}) + 8s \times 1 - 4s^{-3}(4t^3)$$

$$= -2s^3t^{-3} + 8s - 16s^{-3}t^3$$

As with problems in one independent variable, it does not matter which letters are used to represent the variables.

c. $h(x, y, z) = x^3yz^2 + 9y^4z^{-1} + 2x^{\frac{1}{2}}y^5z^2$:

$$\frac{\partial h(x, y, x)}{\partial x} = yz^2 \times \frac{\partial}{\partial x}x^3 + 9y^4z^{-1} \times \frac{\partial}{\partial x}1 + 2y^5z^2 \times \frac{\partial}{\partial x}x^{\frac{1}{2}}$$

$$= yz^2(3x^2) + 9y^4z^{-1}(0) + 2y^5z^2(\frac{1}{2}x^{-\frac{1}{2}}) = 3x^2yz^2 + x^{-\frac{1}{2}}y^5z^2$$

$$\frac{\partial h(x, y, x)}{\partial y} = x^3z^2 \times \frac{\partial}{\partial y}y + 9z^{-1} \times \frac{\partial}{\partial y}y^4 + 2x^{\frac{1}{2}}z^2 \times \frac{\partial}{\partial y}y^5$$

$$= x^3z^2(1) + 9z^{-1}(4y^3) + 2x^{\frac{1}{2}}z^2(5y^4) = x^3z^2 + 36y^3z^{-1} + 10x^{\frac{1}{2}}y^4z^2$$

$$\frac{\partial h(x, y, x)}{\partial z} = x^3y \times \frac{\partial}{\partial z}z^2 + 9y^4 \times \frac{\partial}{\partial z}z^{-1} + 2x^{1/2}y^5 \times \frac{\partial}{\partial z}z^2$$

$$= x^3y(2z) + 9y^4(-z^{-2}) + 2x^{1/2}y^5(2z) = 2x^3yz - 9y^4z^{-2} + 4x^{1/2}y^5z$$

This function had three independent variables, so we needed to calculate three first-order partials.

Problem 4.1.4. Find the first-order partials of the following functions.

a. $f(x, y) = (5x^3 + 7)(8y^2 - 9)$: Think of $g(x, y) = (5x^3 + 7)$ and $h(x, y) = (8y^2 - 9)$ and apply the product rule as follows:

$$\frac{\partial f(x, y)}{\partial x} = \frac{\partial g(x, y)}{\partial x}h(x, y) + g(x, y)\frac{\partial h(x, y)}{\partial x}$$

$$= \frac{\partial(5x^3 + 7)}{\partial x}(8y^2 - 9) + (5x^3 + 7)\frac{\partial(8y^2 - 9)}{\partial x}$$

$$= 15x^2 \times (8y^2 - 9) + (5x^3 + 7) \times 0 = 15x^2(8y^2 - 9)$$

$$\frac{\partial f(x, y)}{\partial y} = \frac{\partial g(x, y)}{\partial y}h(x, y) + g(x, y)\frac{\partial h(x, y)}{\partial y}$$

$$= \frac{\partial(5x^3 + 7)}{\partial y}(8y^2 - 9) + (5x^3 + 7)\frac{\partial(8y^2 - 9)}{\partial y}$$

$$= 0 \times (8y^2 - 9) + (5x^3 + 7) \times 16y = 16y(5x^3 + 7)$$

(*Note:* One need not use the product rule for this problem. For $\partial f/\partial x$, the factor $(8y^2 - 9)$ is just a constant and can be pulled through the differentiation. Similarily, $(5x^3 + 7)$ is considered a constant when looking for $\partial f/\partial y$.)

b. $f(x) = (x^4 - 8y)(7x^2 + y^5)$: Think of $g(x) = (x^4 - 8y)$ and $h(x) = (7x^2 + y^5)$, then:

$$\frac{\partial f(x)}{\partial x} = \frac{\partial(x^4 - 8y)}{\partial x}(7x^2 + y^5) + (x^4 - 8y)\frac{\partial(7x^2 + y^5)}{\partial x}$$

$$= 4x^3 \times (7x^2 + y^5) + (x^4 - 8y) \times 14x = 4x^3(7x^2 + y^5) + 14x(x^4 - 8y)$$

$$= 28x^5 + 4x^3y^5 + 14x^5 - 112xy = 42x^5 + 4x^3y^5 - 112xy$$

$$\frac{\partial f(x)}{\partial y} = \frac{\partial(x^4 - 8y)}{\partial y}(7x^2 + y^5) + (x^4 - 8y)\frac{\partial(7x^2 + y^5)}{\partial y}$$

$$= -8 \times (7x^2 + y^5) + (x^4 - 8y) \times 5y^4 = -8(7x^2 + y^5) + 5y^4(x^4 - 8y)$$

$$= -56x^2 - 8y^5 + 5x^4y^4 - 40y^5 = 5x^4y^4 - 56x^2 - 48y^5$$

c. $f(x, y) = \dfrac{3xy}{2x + 6y}$: Think of $g(x, y) = 3xy$ and $h(x, y) = 2x + 6y$, then:

$$\frac{\partial f(x, y)}{\partial x} = \frac{\dfrac{\partial g(x, y)}{\partial x} h(x, y) - g(x, y) \dfrac{\partial h(x, y)}{\partial x}}{[h(x, y)]^2}$$

$$= \frac{\dfrac{\partial 3xy}{\partial x}(2x + 6y) - 3xy \times \dfrac{\partial (2x + 6y)}{\partial x}}{(2x + 6y)^2}$$

$$= \frac{3y(2x + 6y) - 3xy \times 2}{(2x + 6y)^2} = \frac{(6xy + 18y^2) - 6xy}{(2x + 6y)^2} = \frac{18y^2}{(2x + 6y)^2}$$

$$\frac{\partial f(x, y)}{\partial y} = \frac{\dfrac{\partial g(x, y)}{\partial y} h(x, y) - g(x, y) \dfrac{\partial h(x, y)}{\partial y}}{[h(x, y)]^2}$$

$$= \frac{\dfrac{\partial 3xy}{\partial y}(2x + 6y) - 3xy \times \dfrac{\partial (2x + 6y)}{\partial y}}{(2x + 6y)^2}$$

$$= \frac{3x(2x + 6y) - 3xy \times 6}{(2x + 6y)^2} = \frac{(6x^2 + 18xy) - 18xy}{(2x + 6y)^2} = \frac{6x^2}{(2x + 6y)^2}$$

Problem 4.1.5. Find the first-order partials of the following functions.

a. $f(x, y) = (x^3 + 2y^4)^2$: Think of $g(z) = z^2$ and $h(x, y) = x^3 + 2y^4$, then:

$$\frac{\partial f(x, y)}{\partial x} = \frac{\partial h(x, y)}{\partial x} g'[h(x, y)] = \frac{\partial (x^3 + 2y^4)}{\partial x} \times 2(x^3 + 2y^4)$$
$$= 3x^2 \times 2(x^3 + 2y^4) = 6x^2(x^3 + 2y^4)$$

$$\frac{\partial f(x, y)}{\partial y} = \frac{\partial h(x, y)}{\partial y} g'[h(x, y)] = \frac{\partial (x^3 + 2y^4)}{\partial y} \times 2(x^3 + 2y^4)$$
$$= (2 \times 4y^3) \times 2(x^3 + 2y^4) = 16y^3(x^3 + 2y^4)$$

b. $f(x, y) = (9xy^2 - 3)^{-2}$: Think of $g(z) = z^{-2}$ and $h(x, y) = 9xy^2 - 3$, then:

$$\frac{\partial f(x, y)}{\partial x} = \frac{\partial h(x, y)}{\partial x} g'[h(x, y)] = \frac{\partial (9xy^2 - 3)}{\partial x} \times (-2)(9xy^2 - 3)^{-3}$$
$$= 9y^2 \times (-2)(9xy^2 - 3)^{-3} = -18y^2(9xy^2 - 3)^{-3}$$
$$\frac{\partial f(x, y)}{\partial y} = \frac{\partial h(x, y)}{\partial y} g'[h(x, y)] = \frac{\partial (9xy^2 - 3)}{\partial y} \times (-2)(9xy^2 - 3)^{-3}$$
$$= 18xy \times (-2)(9xy^2 - 3)^{-3} = -36xy(9xy^2 - 3)^{-3}$$

c. $f(x, y) = e^{x^2y^3}$: Think of $g(z) = e^z$ and $h(x, y) = x^2y^3$, then:

$$\frac{\partial f(x, y)}{\partial x} = \frac{\partial h(x, y)}{\partial x} g'[h(x, y)] = \frac{\partial(x^2y^3)}{\partial x} e^{x^2y^3} = 2xy^3e^{x^2y^3}$$

$$\frac{\partial f(x, y)}{\partial y} = \frac{\partial h(x, y)}{\partial y} g'[h(x, y)] = \frac{\partial(x^2y^3)}{\partial y} e^{x^2y^3} = 3x^2y^2e^{x^2y^3}$$

Problem 4.1.6. Find all the second-order partials (including mixed partials) for the following functions.

a. $f(x, y) = x^2 + y^2$: Begin by finding the first-order partials:

$$\frac{\partial f(x, y)}{\partial x} = f_x = 2x \text{ and } \frac{\partial f(x, y)}{\partial y} = f_y = 2y$$

To find the four second-order partials (f_{xx}, f_{xy}, f_{yy}, f_{yx}), simply consider the first-order partials to be two new functions and take the partials of each with respect to x and y:

$$f_{xx} = 2, f_{xy} = 0, f_{yy} = 2, \text{ and } f_{yx} = 0$$

Notice that $f_{xy} = f_{yx}$. This is as expected and will be true for all functions contained in this monograph.

b. $f(x, y) = x^2 + 4y^2 - 4x$: Begin by finding the first-order partials:

$$f_x = 2x - 4 \text{ and } f_y = 8y$$

Now find the four second-order partials:

$$f_{xx} = 2, f_{xy} = 0, f_{yy} = 8, \text{ and } f_{yx} = 0$$

c. $z = x^3 + y^3 - 3xy$: The change from $f(x, y)$ to z makes no difference, and we again begin by finding the first-order partials:

$$f_x = 3x^2 - 3y \text{ and } f_y = 3y^2 - 3x$$

Now find the second-order partials:

$$f_{xx} = 6x, f_{xy} = -3, f_{yy} = 6y, \text{ and } f_{yx} = -3$$

4.2 Extrema of Multivariate Functions

Recall that the second derivative test for multivariate extrema uses the discriminant $D(a, b) = [f_{xx}(a, b) \times f_{yy}(a, b)] - [f_{xy}(a, b)]^2$, where a and b are the x- and y-values of the critical point. If $D(a, b) > 0$ and $f_{xx} < 0$, the point is a local maximum. If $D(a, b) > 0$ and $f_{xx} > 0$, the point is a local minimum. If $D(a, b) < 0$, the point is a saddle point. If $D(a,b) = 0$, the test gives no information about the point.

Problem 4.2.1. Find the high or low point of $f(x, y) = x^2 + y^2$. Recall that for a point to be an extremum, $f_x = 0$ and $f_y = 0$. Thus, we must begin by finding the first-order partials of the function, setting them equal to 0, and solving. From Problem 4.1.6a we know $f_x = 2x$ and $f_y = 2y$. Setting each equal to 0 and solving tells us that $f_x = 2x = 0 \Rightarrow x = 0$ and $f_y = 2y = 0 \Rightarrow y = 0$. Entering these values into the original function yields the critical point $(0, 0, 0)$. By looking at the function, we should be able to guess that the critical point is a minimum. To verify this, use the second-order partials ($f_{xx} = 2$, $f_{xy} = 0$, $f_{yy} = 2$, and $f_{yx} = 0$) to form the discriminant:

$$D(0, 0) = [f_{xx}(0, 0) \times f_{yy}(0, 0)] - [f_{xy}(0, 0)]^2 = (2 \times 2) - 0 = 4$$

Because $D(0, 0) = 4 > 0$ and $f_{xx} = 2 > 0$, the critical point is a minimum. There are two additional things to point out about the discriminant. First, only values for x and y are used (which is why it is denoted as $D(0, 0)$ rather than $D(0, 0, 0)$). Second, recall that the formula for the discriminant makes use of $f_{xy} = f_{yx}$ for f continuous at the critical point.

Problem 4.2.2. Find the high or low point of $f(x, y) = x^2 - y^2$. This function is quite similar to the one in the previous problem. The first-order partials will be $f_x = 2x$ and $f_y = -2y$. Setting the first-order partials equal to 0 and solving again yields the critical point $(0, 0, 0)$. The second-order partials for this function are $f_{xx} = 2$, $f_{xy} = 0$, $f_{yy} = -2$, and $f_{yx} = 0$. Forming the discriminant we find the following:

$$D(0, 0) = [f_{xx}(0, 0) \times f_{yy}(0, 0)] - [f_{xy}(0, 0)]^2 = 2 \times (-2) - 0 = -4$$

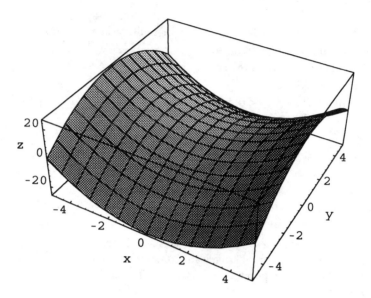

Figure 4.1. Graph for Problem 4.2.2: Saddle Point

Because $D(0, 0) = -4 < 0$, the critical point is not an extremum, but it is a *saddle point*. As you can see from Figure 4.1, the graph of the function around the critical point resembles a saddle. Notice also that if y is held constant, $(0, 0, 0)$ is a minimum in the xz-plane, and if x is held constant, $(0, 0, 0)$ is a maximum in the yz-plane. (*Note:* $z = f(x, y)$.)

Problem 4.2.3. Find the high or low points of $z = x^3 + y^3 - 3xy$. We know from Problem 4.1.6c that the first-order partials are $f_x = 3x^2 - 3y$ and $f_y = 3y^2 - 3x$. Setting them equal to 0 and solving requires a bit more work for this problem. Begin with f_x:

$$f_x = 3x^2 - 3y = 0 \Rightarrow 3x^2 = 3y \Rightarrow x^2 = y$$

Now substitute this value for y into f_y and proceed as follows:

$$f_y = 3y^2 - 3x = 0 \Rightarrow y^2 - x = 0 \Rightarrow (x^2)^2 - x = 0 \Rightarrow x^4 - x = 0 \Rightarrow x^3 - 1 = 0$$
$$\Rightarrow (x - 1)(x^2 + x + 1) = 0 \Rightarrow x = 1$$

Factoring $x^3 - 1$ may not have been obvious, but with a little experimentation you should have found that $x - 1$ was one of its factors. You can use the quadratic formula to see that $x^2 + x + 1$ does not have any real roots, so $x = 1$ is the only real root of $x^3 - 1$. We can use this value of x to find that $y = 1$ and the critical point is $(1, 1, -1)$. Turning to the second-order partials, we find $f_{xx} = 6x^2, f_{xy} = -3, f_{yy} = 6y^2$, and $f_{yx} = -3$. Forming the discriminant yields the following:

$$D(1, 1) = [6(1)^2 \times 6(1)^2] - (-3)^2 = (6 \times 6) - 9 = 36 - 9 = 27$$

Because $D(1, 1) > 0$ and $f_{xx} > 0$, $(1, 1, -1)$ is a minimum. (We also can see from the function that this is a *local* minimum.)

During the above the calculations you may have noticed that $x^4 - x = 0 \Rightarrow x^4 = x$. This suggests that $(0, 0, 0)$ is also a critical point, but $D(0, 0) = 0 + 0 - 9 = -9 < 0$, so $(0, 0, 0)$ is a saddle point.

Problem 4.2.4. A political consultant determines that the function for estimating votes on the basis of money spent is $f(x, y) = x^2 + xy + y^2 - 2x - 6y$, where x and y represent the thousands of dollars spent on television and radio, respectively, and $f(x, y)$ represents thousands of votes. Does this function have any extrema? If the current budget allocates \$40,000 for television and \$15,000 for radio, where should an additional \$1,000 be spent for the best results?

Begin by finding that the first-order partials are $f_x = 2x + y - 2$ and $f_y = x + 2y - 6$. We want to know when both partials are equal to 0, so set them equal to each other and try to get one variable in terms of the other:

$$2x + y - 2 = x + 2y - 6 \Rightarrow x + 4 = y$$

Now substitute this value of y into one of the partials, say f_x:

$$f_x = 2x + y - 2 = 2x + (x + 4) - 2 = 3x + 2 = 0$$
$$\Rightarrow 3x = -2 \Rightarrow x = \frac{-2}{3}$$

Using this value we find $y = x + 4 = \frac{-2}{3} + 4 = \frac{10}{3}$. Entering these values into the original equation, we find the critical point to be

$\left(\dfrac{-2}{3}, \dfrac{10}{3}, \dfrac{-28}{3}\right)$. This critical point suggests the possibility of negative spending on television and the obtaining of negative votes. Clearly, the feasible set would not allow this, but let us continue to see whether the critical point is a minimum. The second-order partials are $f_{xx} = 2, f_{xy} = 1, f_{yy} = 2$, and $f_{xy} = 1$. Forming the discriminant we find $D\left(\dfrac{-2}{3}, \dfrac{10}{3}\right) = (2 \times 2) - 1^2 = 4 - 1 = 3$. Because both $D\left(\dfrac{-2}{3}, \dfrac{10}{3}\right)$ and f_{xx} are positive, $\left(\dfrac{-2}{3}, \dfrac{10}{3}, \dfrac{-28}{3}\right)$ is a minimum. (As a practical matter, however, the feasible set requires $x \geq 0$ and $y \geq 0$.)

We can make use of the first partials to determine where to spend the additional \$1,000. The first partials tell us the slope of the tangent line with respect to each axis at a given point, so the additional \$1,000 should be spent on television time if $f_x(40, 15) > f_y(40, 15)$. The extra money should be spent on radio time if the reverse is true. Entering these values into the first partials, we find the following:

$$f_x(40, 15) = 2x + y - 2 = 2(40) + 15 - 2 = 93$$
$$f_y(40, 15) = x + 2y - 6 = 40 + 2(15) - 6 = 64$$

Thus, the money should be spent on television advertising to obtain the most additional votes. You can verify this for yourself by calculating the values of $f(40, 15) = 2255$, $f(41, 15) = 2349$, and $f(40, 16) = 2320$ from the original function.

Problem 4.2.5. Consider situation of Problem 4.2.4 again, but suppose that the total budget for television and radio advertising will not be less than \$60,000. Given this budget constraint, what is the minimum number of votes the candidate could receive?

In essence, the problem is to minimize $f(x, y) = x^2 + xy + y^2 - 2x - 6y$ subject to the constraint $g(x, y) = x + y - 60 = 0$. We will use the method of Lagrange multipliers to solve this problem. Begin by constructing the following function:

$$F(x, y, \lambda) = f(x, y) - \lambda \times g(x, y) = x^2 + xy + y^2 - 2x - 6y - \lambda(x + y - 60)$$

By constructing this new function $F(x, y, \lambda)$, we limit our search for extrema to only those values of x and y that satisfy the constraint. We again begin by finding the first-order partials $F_x = 2x + y - 2 - \lambda$, $F_y = x + 2y - 6 - \lambda$, and $F_\lambda = -x - y + 60$. Setting each of these equal to 0 and solving give us three equations and three unknowns. It might be easiest to start with F_λ and note that

$$F_\lambda = -x - y + 60 = 0 \Rightarrow x = 60 - y$$

Enter this value for x back into F_x and F_y to find

$$F_x = 2x + y - 2 - \lambda = 2(60 - y) + y - 2 - \lambda$$
$$= 120 - 2y + y - 2 - \lambda = 118 - y - \lambda = 0$$

$$F_y = x + 2y - 6 - \lambda = (60 - y) + 2y - 6 - \lambda$$
$$= 60 - y + 2y - 6 - \lambda = 54 + y - \lambda = 0$$

Now we have two equations and two unknowns. Next, set the two equations equal to each other and solve

$$118 - y - \lambda = 0 = 54 + y - \lambda \Rightarrow 64 = 2y \Rightarrow y = 32$$

Thus, $x = 60 - y = 60 - 32 = 28$, and you can use either F_x or F_y to find that $\lambda = 86$. Entering these values into $F(x, y, \lambda)$, we find

$$
\begin{aligned}
F(x, y, \lambda) &= x^2 + xy + y^2 - 2x - 6y - \lambda(x + y - 60) \\
&= (28)^2 + (28)(32) + (32)^2 - 2(28) - 6(32) - 86(28 + 32 - 60) \\
&= 784 + 896 + 1024 - 56 - 192 \\
&= 2456
\end{aligned}
$$

We will come back to the use of Lagrange multipliers in the problems for Chapter 6, so I will forgo using a form of the second derivative test to verify that $(28, 32, 2456)$ is a minimum given the constraint. We can, however, enter values for x and y near the critical point to see whether $f(x, y)$ will be more than $f(28, 32)$, but remember that the new values for x and y must conform to the constraint. For example, $f(29, 31) = 2457$ and $f(27, 33) = 2457$.

Problem 4.2.6. Suppose the Internal Revenue Service (IRS) is trying a new method of checking tax returns. The director of the IRS must budget x thousand dollars for employee salaries and y thousand dollars for new equipment, with the expectation that $f(x, y) = 60x^{1/3}y^{2/3}$ tax returns can be processed. If the director's budget is \$120,000, how should the money be allocated to maximize the number of returns checked?

The constraint is $g(x, y) = x + y - 120 = 0$, so construct $F(x, y, \lambda)$ as:

$$F(x, y, \lambda) = f(x, y) - \lambda \times g(x, y) = 60x^{1/3}y^{2/3} - \lambda(x + y - 120)$$

Now find the first-order partials of $F(x, y, \lambda)$:

$$F_x = \left(\frac{1}{3}\right)60x^{-2/3}y^{2/3} - \lambda = 20x^{-2/3}y^{2/3} - \lambda = 0$$

$$F_y = \left(\frac{2}{3}\right)60x^{1/3}y^{-1/3} - \lambda = 40x^{1/3}y^{-1/3} - \lambda = 0$$

$$F_\lambda = -x - y + 120 = 0$$

To solve this system of equations, begin by setting $F_x = F_y$ and simplifying

$$20x^{-2/3}y^{2/3} - \lambda = 40x^{1/3}y^{-1/3} - \lambda \Rightarrow x^{-2/3}y^{2/3} = 2x^{1/3}y^{-1/3} \Rightarrow y^{2/3} = 2xy^{-1/3} \Rightarrow y = 2x$$

Substituting this result into F_λ yields

$$F_\lambda = -x - y + 120 = -x - 2x + 120 = -3x + 120 = 0 \Rightarrow 3x = 120 \Rightarrow x = 40$$

Entering this value back into F_λ tells us that $y = 80$. Entering these values into $f(x, y)$ yields a critical point of (40, 80, 3809.76). As with the previous problem, to verify that this is a minimum, we must check values near the critical point that satisfy the constraint. For example, $f(39, 81) = 3809.16$ and $f(41, 79) = 3809.17$. Thus, if \$40,000 is spent on employee salaries and \$80,000 is spent on new equipment, the director can expect that about 3.8 million tax returns will be checked using the new method.

Problem 4.2.7. An open rectangular box has a surface area of A. What relative dimensions of the box will maximize the volume of the box?

Here we must construct both the constraint and the function we wish to maximize. The function for the volume of the box is the length (l) times the width (w) times the height (h). Thus, $V(l, w, h) = lwh$. Constructing the constraint, which is the area A, will take a bit more work. Because the box is open, we need to find the surface area of the bottom (lw) plus two sides ($2lh$) plus another two sides ($2wh$). Thus, $A(l, w, h) = lw + 2lh + 2wh$. Unlike previous problems, we do not know the precise value of the constraint. Instead, let us call the area k so we can put the constraint in the proper form $A(l, w, h) = lw + 2lh + 2wh - k = 0$. Now construct $F(l, w, h, \lambda)$ as:

$$F(l, w, h, \lambda) = V(l, w, h) - \lambda \times A(l, w, h) = lwh - \lambda(lw + 2lh + 2wh - k)$$

Now find the first-order partials of $F(l, w, h, \lambda)$:

$$F_l = wh - \lambda w - 2\lambda h = 0 \qquad F_h = lw - 2\lambda l - 2\lambda w = 0$$
$$F_w = lh - \lambda l - 2\lambda h = 0 \qquad F_\lambda = -lw - 2lh - 2wh + k = 0$$

Though this system of equations appears formidable, the solution is not too difficult. Begin by setting F_l and F_w equal to one another and solving:

$$wh - \lambda w - 2\lambda h = lh - \lambda l - 2\lambda h \Rightarrow wh - \lambda w = lh - \lambda l \Rightarrow w(h - \lambda)$$
$$= l(h - \lambda) \Rightarrow w = l$$

Now substitute l for w in F_h and solve for λ:

$$F_h = lw - 2\lambda l - 2\lambda w = l^2 - 2\lambda l - 2\lambda l = l^2 - 4\lambda l = 0$$
$$\Rightarrow l^2 = 4\lambda l \Rightarrow l = 4\lambda \Rightarrow \lambda = \frac{l}{4}$$

Now that we have both w and λ in terms of l, use F_w to solve for h:

$$F_w = lh - \left(\frac{l}{4}\right)l - 2\left(\frac{l}{4}\right)h = lh - \frac{l^2}{4} - \frac{l}{2}h = \frac{l}{2}h - \frac{l^2}{4} = 0$$
$$\Rightarrow \frac{l}{2}h = \frac{l^2}{4} \Rightarrow lh = \frac{l^2}{2} \Rightarrow h = \frac{l}{2}$$

Thus, to maximize the volume of an open box, we must select the dimensions such that the length and width are equal and the height is half the length. (Check for yourself that this is a maximum by using some actual values.)

4.3 Homework Problems

1. Find the first-order partials of the following functions.

   a. $f(x, y) = 3x^4 - 9y$

   e. $f(x, y) = \dfrac{x - y}{x^2 - y^2}$

   b. $f(x, y, z) = 7x^3y^2z^7$

   f. $f(r, s) = (4r^2s^4 - 9)^3$

   c. $f(s, t) = 8s^3t + 5s^2t^2 - st^3$

   g. $g(x, y) = e^{2xy}$

   d. $h(s, t) = (6s - 4t^3)(1 - t)$

   h. $f(x, y) = e^{x^2 - 9y^2}$

2. Find all the first- and second-order partials of the following functions.

   a. $f(x, y) = 2x^3 - 7y^2$

   c. $f(u, v) = u^4 - u^2v - uv^3$

   b. $f(s, t) = 9s^3t - 3st^2 + 4t^3$

   d. $z = x^7y^2 - 6xy$

3. Find the extrema of the following functions. Form the discriminant to verify your answer.

   a. $f(x, y) = x^2 + xy + y^2 + 3x - 3y + 4$

   b. $f(x, y) = 6x^2 + 2y^2 + 6xy + 6x - 12y + 7$

   c. $f(r, s) = 6rs - 7r^2 - 2s^2 + 4r - 6s + 2$

   d. $f(u, v) = u^2 + uv + 3u + 2v + 5$

4. Minimize $4x^2 + 3y^2$ subject to the constraint $8x - 3y = 19$.

5. Maximize xyz subject to the constraint $x + y + z = N$.

6. An agency field office provides two informational booklets to the public for a small fee. One of the booklets is locally produced and costs the office 30 cents per booklet. The other is produced by the agency headquarters and costs the office 40 cents per booklet. The office director estimates that if the locally produced booklets are sold for x cents each and the booklets produced by headquarters are sold for y cents each, then approximately $70 - 5x + 4y$ of the locally produced booklets and $80 + 6x - 7y$ of the booklets produced by headquarters will be sold each day. How should the office director set the prices of the two booklets to maximize the profit from the booklet sales? (*Hint:* Remember that total profit will be the difference between price of the booklets and their cost multiplied by how many are sold.)

7. A closed rectangular box has a volume of V. What relative dimensions of the box will minimize the surface area of the box? (*Hint: V* is the constraint.)

8. Suppose the material for the bottom of a closed rectangular box costs twice as much as the material for the top and sides. What are the most economical dimensions for a box of a given volume, V? (*Hint:* What is the relationship between A, the area, and C, the cost?)

5. INTEGRAL CALCULUS

5.1 Integration Rules

Recall the basic rules for integration:

1. $\int x^n \, dx = \dfrac{x^{n+1}}{n+1} + c$: This is the power rule for integration.

2. $\int x^{-1} \, dx = \ln |x| + c$: This rule fills the gap in the power rule for x^{-1} (which would not work because there would be a 0 in the denominator). Notice that we take the natural log of the *absolute value* of x because we cannot take the log, natural or otherwise, of a negative number.

3. $\int dx = x + c$: This is actually a special case of the power rule because $dx = 1 \, dx = x^0 \, dx$.

4. $\int [a \times f(x)] \, dx = a \int f(x) \, dx$: You can pull a constant through the integration.

5. $\int [f(x) + g(x)] \, dx = \int f(x) \, dx + \int g(x) \, dx$: The integral of the sum of two functions is equal to the sum of their integrals.

6. $\int f[g(x)] \times g'(x) \, dx = F[g(x)] + c$: This is the chain rule for integration.

7. $\int [g(x) \times h(x)] \, dx = g(x) \times H(x) - \int [g'(x) \times H(x)] \, dx + c,$ where $H(x)$ $= \int h(x) \, dx$: This is the formula for the technique known as integration by parts.

8. $\int e^x \, dx = e^x + c; \int a^x \, dx = \dfrac{a^x}{\ln a} + c$: These rules show the reverse of differentiation rule 9 from Chapter 3.

Note: Recall that in working with integrals, we use the capital of the function letter to designate the integral of the function. Thus, for a function f we use F to indicate the integral of f and f' to indicate the derivative of f.

Problem 5.1.1. Evaluate the following integrals.

a. $\int 6 \, dx$: This is a constant function, and we can apply the power rule as follows:

$$\int 6 \, dx = \int 6x^0 \, dx = 6 \int x^0 \, dx = 6 \left(\frac{x^{0+1}}{0+1} + c_1 \right) = 6 \left(\frac{x^1}{1} + c_1 \right) = 6x + c_2$$

I used rule 4 to pull the constant through the integration and was left with just $\int dx$, which, from rule 3, is a special case of the power rule. Remember to include the constant of the integration, c, when calculating an indefinite integral. (I have shown all the steps in the solution to this problem but will combine the most elementary steps in the following problems.)

b. $\int \dfrac{9}{x} \, dx = \int 9x^{-1} \, dx = 9 \times \ln |x| + c$: This problem makes use of rule 2, which fills in the gap in the power rule for exponents to the -1 power.

c. $\int 5x^3 \, dx = \dfrac{5}{4}x^4 + c$: Again, I used rule 4 to pull the constant through the integration and then applied the power rule.

d. $\int (6x^2 + 7) \, dx = \int 6x^2 \, dx + \int 7 \, dx = 2x^3 + c_1 + 7x + c_2 = 2x^3 + 7x + c_3$: For this problem, I used rule 5 to separate the two terms of the expression into two integrals. Notice that by doing so, I obtained a constant of integration for *each* integration. These two constants were combined for the final expression. Although it is good practice to distinguish between the various constants of integration, it is not all that important. We do not know the value of the constants of integration, and if you had done the integration in one step (basically applying rule 5 in your head), you would have only one constant of integration, so we need not devote too much time to them.

e. $\int \frac{8}{y^3} dy = \int 8y^{-3} dy = -4y^{-2} + c$: Two things are worth noting about this problem. First, be careful when integrating variables with negative exponents. Do not forget that in adding 1 to the negative exponent, you actually are *reducing* its absolute value. (This may seem simple, but I have had many answers of $-2y^{-4} + c$ for this problem.) Second, notice the change in the variable. It does not matter that the expression is in the variable y, *provided* that you are doing the integration with respect to y (i.e., dy rather than dx appears on the right). If dx had appeared to the right, you would do the integration with respect to x, and all appearances of other variables would be treated as constants. For example, if the original problem had asked us to integrate with respect to x, then we would have had $\int 8y^{-3} dx$

$= 8xy^{-3} + c$.

f. $\int \sqrt[4]{x} \, dx = \int x^{1/4} dx = \frac{x^{5/4}}{\frac{5}{4}} + c = \frac{4}{5} x^{5/4} + c$: This is another application of the power rule. Take care in dealing with fractional exponents. Notice that the fraction 5/4 appearing in the denominator becomes 4/5 when the expression is simplified.

g. $\int (t^{-4/3} - 2t^{-1}) \, dt = \int t^{-4/3} dt - \int 2t^{-1} dt = \frac{t^{-1/3}}{-\frac{1}{3}} + c_1 - 2 \ln |t| + c_2 = -3t^{-1/3} - 2 \ln |t|$

$+ c_3$: When the integrand becomes more complex, it is a good idea to write out more of the steps to avoid making simple addition or multiplication mistakes.

h. $\int z^3(z + 1)^2 \, dz = \int z^3(z^2 + 2z + 1)dz = \int (z^5 + 2z^4 + z^3) \, dz = \frac{z^6}{6} + \frac{2z^5}{5} + \frac{z^4}{4} + c$:
Although we might have attempted to evaluate the expression using a more complex technique, do not overlook the obvious. A few simple multiplications allow us to do three easy integrations (one for each term) rather than one difficult one.

Problem 5.1.2. Use the chain rule to evaluate the following indefinite integrals.

a. $\int (5x - 8)^2 \, dx$: Recall that the chain rule for integration uses a technique known as substitution. (Some authors of calculus texts call this *change of variable integration.*) Begin by thinking of $u = 5x - 8$, then $\dfrac{du}{dx} = 5 \Rightarrow du$

$= 5\,dx \Rightarrow dx = \dfrac{du}{5}$. Now make these substitutions:

$$\int (5x - 8)^2\, dx = \int u^2 \frac{du}{5} = \int \frac{u^2}{5}\, du$$

Now you can either pull the constant through the integration or do the integration directly using the power rule:

$$\int \frac{u^2}{5}\, du = \frac{u^3}{5 \times 3} + c = \frac{u^3}{15} + c$$

To complete the problem, reverse the earlier substitution to obtain an answer in terms of x:

$$\frac{u^3}{15} + c = \frac{(5x - 8)^3}{15} + c$$

b. $\int (e^3 - e^{3t})dt$. Begin by applying rule 5 to separate the two terms:

$$\int (e^3 - e^{3t})dt = \int e^3 dt - \int e^{3t} dt$$

The first integral is quite easy because e is just a number (about 2.718), so e raised to a constant is just a constant. Thus,

$$\int e^3 dt = e^3 \int dt = e^3 \times (t + c_1) = e^3 t + c_2$$

The second integral requires use of the chain rule. Think of $u = 3t$, then $\dfrac{du}{dt} = 3 \Rightarrow dt = \dfrac{du}{3}$. Making these substitutions, we find:

$$\int e^{3t} dt = \int e^u \frac{du}{3} = \frac{1}{3} \int e^u\, du = \frac{1}{3}(e^u + c_3) = \frac{e^u}{3} + c_4$$

Now reverse the substitution and combine the two parts:

$$(e^3 t + c_2) - \left(\frac{e^{3t}}{3} + c_4 \right) = e^3 t - \frac{e^{3t}}{3} + c_5$$

c. $\int 2x(x^2 + 1)^2 dx$: We could solve this problem by doing the multiplications and then integrating, but let us use the chain rule. Think of $u = x^2 + 1$, then $\frac{du}{dx} = 2x \Rightarrow du = 2x\,dx$. Making these substitutions, we have

$$\int 2x(x^2 + 1)^2 dx = \int (x^2 + 1)^2\, 2x\,dx = \int u^2 du$$

Notice that for this substitution I used $du = 2x\,dx$ rather than $dx = \frac{du}{2x}$. For the constant, this was just a matter of convenience. The variables u and x, however, must be on opposite sides of the equation. Now perform the integration and reverse the substitutions:

$$\int u^2 du = \frac{u^3}{3} + c = \frac{(x^2 + 1)^3}{3} + c$$

d. $\int \frac{z\,dz}{(z^2 + 1)^3}$: This integral also requires the chain rule. Think of $u = z^2 + 1$, then $\frac{du}{dz} = 2z \Rightarrow du = 2z\,dz \Rightarrow \frac{du}{2} = z\,dz$. Making these substitutions, we have

$$\int \frac{z\,dz}{(z^2 + 1)^3} = \int (z^2 + 1)^{-3}\, z\,dz = \int u^{-3}\frac{du}{2} = \frac{1}{2}\int u^{-3}du$$

Doing the integration and reversing the substitution yields

$$\frac{1}{2}\int u^{-3}du = \frac{1}{2}\left(\frac{u^{-2}}{-2} + c_1\right) = \frac{u^{-2}}{-4} + c_2 = \frac{(z^2 + 1)^{-2}}{-4} + c_2$$

e. $\int x^2(1 + x)^{\frac{1}{2}}dx$: For this problem, choose $u = (1 + x)^{\frac{1}{2}}$, then $u^2 = 1 + x \Rightarrow x = u^2 - 1$ and $dx = 2u\,du$. Making these substitutions and completing the problem yields

$$\int x^2(1 + x)^{\frac{1}{2}}dx = \int (u^2 - 1)^2 u \times 2u\,du = \int 2u^2(u^2 - 1)^2 du = \int 2u^2(u^4 - 2u^2 + 1)\,du$$

$$= \int (2u^6 - 4u^4 + 2u^2)\,du = \frac{2}{7}u^7 - \frac{4}{5}u^5 + \frac{2}{3}u^3 + c$$

$$= \frac{2}{7}[(1+x)^{\frac{1}{2}}]^7 - \frac{4}{5}[(1+x)^{\frac{1}{2}}]^5 + \frac{2}{3}[(1+x)^{\frac{1}{2}}]^3 + c$$

$$= \frac{2}{7}(1+x)^{\frac{7}{2}} - \frac{4}{5}(1+x)^{\frac{5}{2}} + \frac{2}{3}(1+x)^{\frac{3}{2}} + c$$

The key to this problem is making the correct choice for u. This particular choice, however, is not at all obvious. Although we often need to manipulate functions to get them into a form we know how to integrate, this particular technique is one that most people will not consider without having seen it at least once.

Problem 5.1.3. Use integration by parts to evaluate the following integrals.

a. $\int xe^{-x}dx$. Choose $g(x) = x$ and $h(x) = e^{-x}$, then $g'(x) = 1$ and $H(x) = \int h(x)dx = -e^{-x} + c_1$. Now construct the formula for integration by parts:

$$\int [g(x) \times h(x)] \, dx = g(x) \times H(x) - \int [g'(x) \times H(x)] \, dx + c$$

$$\int xe^{-x}dx = x \times (-e^{-x} + c_1) - \int [1 \times (-e^{-x} + c_1)] \, dx$$

$$= -xe^{-x} + c_1 x - \int (-e^{-x} + c_1) \, dx$$

We are left with a second integral, but it is easier to evaluate than the first one:

$$-xe^{-x} + c_1 x - \int (-e^{-x} + c_1) \, dx = -xe^{-x} + c_1 x - (e^{-x} + c_1 x + c_2)$$

$$= -xe^{-x} + c_1 x - e^{-x} - c_1 x + c_3$$

$$= -xe^{-x} - e^{-x} + c_3 = (-x - 1)e^{-x} + c_3$$

(*Note:* The c in the general formula for integration by parts becomes the subscripted constants of integration when performing the integration.)

b. $\int x^2 e^x dx$. Choose $g(x) = x^2$ and $h(x) = e^x$, then $g'(x) = 2x$ and $H(x) = e^x + c_1$. Now enter these values into the formula and do the integration:

$$\int [g(x) \times h(x)] \, dx = g(x) \times H(x) - \int [g'(x) \times H(x)] \, dx + c$$

$$\int x^2 e^x dx = x^2 \times (e^x + c_1) - \int [2x \times (e^x + c_1)] \, dx$$

$$= x^2 e^x + c_1 x^2 - \int (2xe^x + 2c_1 x)\, dx$$

$$= x^2 e^x + c_1 x^2 - \int 2xe^x\, dx - \int 2c_1 x\, dx$$

$$= x^2 e^x + c_1 x^2 - c_1 x^2 + c_2 - \int 2xe^x\, dx$$

$$= x^2 e^x + c_2 - \int 2xe^x\, dx$$

As you can see, we are left with another integral that is "improved" in the sense that x is to a smaller power, but we need to apply integration by parts again. This time, choose $g(x) = 2x$ and $h(x) = e^x$, then $g'(x) = 2$ and $H(x) = e^x + c_3$:

$$\int [g(x) \times h(x)]\, dx = g(x) \times H(x) - \int [g'(x) \times H(x)]\, dx + c$$

$$\int 2xe^x\, dx = 2x \times (e^x + c_3) - \int [2 \times (e^x + c_3)]\, dx$$

$$= 2xe^x + 2c_3 x - \int (2e^x + 2c_3)\, dx$$

$$= 2xe^x + 2c_3 x - \int 2e^x dx - \int 2c_3\, dx$$

$$= 2xe^x + 2c_3 x - 2c_3 x + c_4 - 2e^x + c_5$$

$$= 2xe^x - 2e^x + c_6$$

Now combine the two parts to obtain a final answer:

$$x^2 e^x + c_2 - (2xe^x - 2e^x + c_6) = x^2 e^x - 2xe^x + 2e^x + c_7$$

c. $\int x(x + 3)^{\frac{1}{2}}\, dx$: Choose $g(x) = x$ and $h(x) = (x + 3)^{\frac{1}{2}}$, then $g'(x) = 1$ and $H(x) = \frac{2}{3}(x + 3)^{\frac{3}{2}} + c_1$. Entering these values into the formula, we find

$$\int [g(x) \times h(x)]\, dx = g(x) \times H(x) - \int [g'(x) \times H(x)]\, dx + c$$

$$\int x(x + 3)^{\frac{1}{2}}\, dx = x \times \left(\frac{2}{3}(x + 3)^{\frac{3}{2}} + c_1\right) - \int \left[1 \times \left(\frac{2}{3}(x + 3)^{\frac{3}{2}} + c_1\right)\right] dx$$

$$= \frac{2}{3}x(x + 3)^{\frac{3}{2}} + c_1 x - \left[\frac{2}{3}\left(\frac{2}{5}(x + 3)^{\frac{5}{2}} + c^2\right) + c_1 x\right] + c_3$$

$$= \frac{2}{3}x(x + 3)^{\frac{3}{2}} + c_1 x - c_1 x - \frac{4}{15}(x + 3)^{\frac{5}{2}} - \frac{2}{3}c_2 + c_3$$

$$= \frac{2}{3}x(x + 3)^{\frac{3}{2}} - \frac{4}{15}(x + 3)^{\frac{5}{2}} + c_4$$

5.2 Definite Integrals

Recall the fundamental theorem of calculus: $\int_a^b f(x)dx = F(x)\big|_a^b$ $= F(b) - F(a)$. Recall also these rules and results for definite integrals:

1. If a function is continuous on $[a, b]$, then it is integrable on $[a, b]$.

2. $\int_a^b f(x)\, dx = -\int_a^b f(x)\, dx$: As noted, changing the direction changes the sign.

3. $\int_a^a f(x)dx = 0$.

4. $\int_a^b dx = b - a$: This is the area under the constant function $f(x) = 1$.

5. $\int_a^b [f(x) + g(x)]\, dx = \int_a^c f(x)\, dx + \int_c^b g(x)\, dx$: This tells us that rule 5 for indefinite integrals applies to definite integrals as well.

6. For $a < c < b$, $\int_a^b f(x)\, dx = \int_a^b f(x)\, dx + \int_a^b f(x)\, dx$: The area under a curve for an interval is equal to the area under the curve for the sum of the subintervals. (This rule is also true for $a \leq c \leq b$, because of rule 3.)

7. For a function f, the average value of f on $[a, b]$ is $\dfrac{\int_a^b f(x)\, dx}{b - a}$.

Problem 5.2.1. Find the area under the curve for these definite integrals.

a. $\int_0^3 3\, dx$: This is a constant function. We can pull the constant through the integration and make use of rule 4 as follows:

$$\int_0^3 3\, dx = 3 \int_0^3 dx = 3x \big|_0^3 = 3(3 - 0) = 9$$

b. $\int_{-1}^2 (x + 6)\, dx$: For this linear function, use rule 5 to separate the two terms, then integrate them separately:

$$\int_{-1}^{2} (x + 6)\, dx = \int_{-1}^{2} x\, dx + \int_{-1}^{2} 6\, dx = \frac{x^2}{2}\Big|_{-1}^{2} + 6x\Big|_{-1}^{2}$$

$$= \left(\frac{2^2}{2} - \frac{(-1)^2}{2}\right) + [(6 \times 2) - 6 \times (-1)]$$

$$= \left(2 - \frac{1}{2}\right) + (12 + 6) = \frac{3}{2} + 18 = \frac{3 + 36}{2} = \frac{39}{2}$$

c. $\int_{-3}^{2} (4t^2 + 4t - 24)\, dt$: This problem again makes use of rule 5:

$$\int_{-3}^{2} (4t^2 + 4t - 24)dt = \left(\frac{4}{3}t^3 + 2t^2 - 24t\right)\Big|_{-3}^{2}$$

$$= \left(\frac{4}{3}(2)^3 + 2(2)^2 - 24(2)\right) - \left(\frac{4}{3}(-3)^3 + 2(-3)^2 - 24(-3)\right)$$

$$= \left(\frac{32}{3} + 8 - 48\right) - (-36 + 18 + 72)$$

There are two things to note about this problem. First, notice that I shortened the process by applying the fundamental theorem of calculus to all three terms at once rather than for each term individually. Second, in evaluating a definite integral, we are looking for the area under a curve for a specified interval, and you may have wondered about the negative value obtained for this integral. You should recognize that $4t^2 + 4t - 24$ will be parabolic. If you differentiate the function, you will see that it has an absolute minimum at $\left(\frac{-1}{2}, -25\right)$. If you factor the function, you will see that it crosses the t-axis at 2 and -3. Thus, for the chosen interval, the function is *below* the t-axis, which we define as *negative* area.

d. $\int_{0}^{2} 2e^t\, dt$: Although this function involves e, we apply the rules as usual:

$$\int_{0}^{2} 2e^t\, dt = 2e^t \Big|_{0}^{2} = 2e^2 - 2e^0 = 2e^2 - 2 \approx 12.78$$

Depending on the context of the problem, you may not want to convert e^2 to its numerical form, but you should recognize that $e^0 = 1$.

e. $\int_4^9 \dfrac{(4x^{\frac{1}{2}}+2)^3}{x^{\frac{1}{2}}}\,dx$: Use the chain rule to do this integration. Let $u = 4x^{\frac{1}{2}} +$

2, then $\dfrac{du}{dx} = 4 \times \dfrac{1}{2}x^{-\frac{1}{2}} = 2x^{-\frac{1}{2}} \Rightarrow \dfrac{du}{2} = x^{-\frac{1}{2}}\,dx$. Because $\dfrac{1}{x^{\frac{1}{2}}} = x^{-\frac{1}{2}}$, rewrite

the function using $x^{-\frac{1}{2}}$ and make these substitutions:

$$\int_4^9 \dfrac{(4x^{\frac{1}{2}}+2)^3}{x^{\frac{1}{2}}}\,dx = \int_4^9 (4x^{\frac{1}{2}}+2)^3 \, x^{-\frac{1}{2}}\,dx = \int_4^9 u^3\,\dfrac{du}{2} = \int_4^9 \dfrac{1}{2}u^3\,du = \dfrac{1}{2}\times\dfrac{1}{4}u^4\,\Big|_4^9$$

$$= \dfrac{1}{8}(4x^{\frac{1}{2}}+2)^4\,\Big|_4^9 = \left(\dfrac{1}{8}[4(9)^{\frac{1}{2}}+2]^4\right) - \left(\dfrac{1}{8}[4(4)^{\frac{1}{2}}+2]^4\right)$$

$$= \left(\dfrac{1}{8}(12+2)^4\right) - \left(\dfrac{1}{8}(8+2)^4\right) = \left(\dfrac{1}{8}(14)^4\right) - \left(\dfrac{1}{8}(10)^4\right)$$

$$= \dfrac{1}{8}(38{,}416) - \dfrac{1}{8}(10{,}000) = \dfrac{1}{8}(28{,}416) = 3552$$

Note: To be more precise, the limits of integration (here 4 and 9) should change when the variable is changed from x to u. Some authors of calculus texts indicate the change using $\int_{u(a)}^{u(b)}$, where a and b are the limits of integration. Other authors will recalculate the values in terms of u (e.g., in this problem $u(4) = 4(4)^{\frac{1}{2}} + 2 = 10$ and $u(9) = 4(9)^{\frac{1}{2}} + 2 = 14$). Some authors sidestep the problem by treating the change-of-variable portion of the process as an indefinite integral. Because the change of variable is temporary, I have chosen to leave the limits of integration as they originally appear. Although not technically correct, it makes for one less thing to worry about in these problems.

f. $\int_0^1 t^2 e^{3t^3}\,dt$: Let $u = 3t^3$, then $\dfrac{du}{dt} = 9t^2 \Rightarrow \dfrac{du}{9} = t^2\,dt$. Now make the substitutions and do the integration:

$$\int_0^1 t^2 e^{3t^3}\,dt = \int_0^1 e^u\,\dfrac{du}{9} = \dfrac{1}{9}\int_0^1 e^u\,du = \dfrac{1}{9}\left(e^u\,\Big|_0^1\right) = \dfrac{1}{9}\left(e^{3t^3}\,\Big|_0^1\right) = \dfrac{1}{9}(e^3 - e^0) \approx 2.12$$

Notice that for this problem I pulled the constant through the integration to make the calculations a bit easier.

g. $\int_1^2 te^{2t}\,dt$: This problem requires integration by parts. Choose $g(t) = t$ and $h(t) = e^{2t}$, then $g'(t) = 1$ and $H(t) = \frac{1}{2}e^{2t}$. Now enter these values into the formula:

$$\int_a^b [g(t) \times h(t)]\,dt = g(t) \times H(t) - \int_a^b [g'(t) \times H(t)]\,dt$$

$$\int_1^2 te^{2t}dt = t \times \left(\frac{1}{2}e^{2t}\right)\Big|_1^2 - \int_1^2 \left[1 \times \left(\frac{1}{2}e^{2t}\right)\right]dt$$

$$= \frac{1}{2}te^{2t}\Big|_1^2 - \int_1^2 \frac{1}{2}e^{2t}dt = \frac{1}{2}te^{2t}\Big|_1^2 - \frac{1}{4}e^{2t}\Big|_1^2$$

$$= \left(\frac{1}{2}te^{2t} - \frac{1}{4}e^{2t}\right)\Big|_1^2 = \left(\frac{1}{2}(2)e^{2(2)} - \frac{1}{4}e^{2(2)}\right) - \left(\frac{1}{2}(1)e^{2(1)} - \frac{1}{4}e^{2(1)}\right)$$

$$= \left(e^4 - \frac{1}{4}e^4\right) - \left(\frac{1}{2}e^2 - \frac{1}{4}e^2\right) = \frac{3}{4}e^4 - \frac{1}{4}e^2 \approx 39.10$$

Problem 5.2.2. A population expert determines that x months from now the population of a particular county will be increasing at the rate of $9 + 5x^{2/3}$ people per month. By how much will the population of the county increase in the next year?

We are being asked to calculate the area under the curve $9 + 5x^{2/3}$ between 0 and 12 (i.e., $\int_0^{12} 9 + 5x^{2/3}\,dx$). This is a fairly simple integration, and we proceed as follows:

$$\int_0^{12} 9 + 5x^{2/3}\,dx+ = \left[9x + 5\left(\frac{3}{5}\right)x^{5/3}\right]\Big|_0^{12} = (9x + 3x^{5/3})\Big|_0^{12}$$

$$= [9(12) + 3(12)^{5/3}] - [9(0) + 3(0)^{5/3}]$$

$$= [9(12) + 3(12)^{5/3}] \approx 108 + 3(62.9) \approx 108 + 188.7$$

$$\approx 296.7$$

Problem 5.2.3. A labor expert calculates that the probability density function for workers who have been laid off and remain unemployed for a particular state is $f(t) = \frac{1}{2}e^{-t/2}$, where $t > 0$ and represents the number of months of unemployment of a randomly selected worker. What is the probability that a randomly selected unemployed worker has been unemployed for 4 months or less? What is the probability that a randomly selected unemployed worker has been unemployed 12 months or more?

Recall that we use probability density functions to determine the probability that an observed value falls within a specified range rather than the probability of a specific outcome. Recall also that probability density functions must satisfy two conditions: (a) $f(x) \geq 0$ for all x, and (b) $\int_{-\infty}^{+\infty} f(x)\, dx = 1$. The first question asks us to find the area under the function between 0 and 4 months (i.e., $\int_{0}^{4} \frac{1}{2} e^{-t/2}\, dt$). Using the chain rule to do the integration, let $u = \frac{-1}{2} t$, then $\frac{du}{dt} = \frac{-1}{2} \Rightarrow -2du = dt$. Proceding as usual:

$$\int_{0}^{4} \frac{1}{2} e^{-t/2}\, dt = \int_{0}^{4} e^{u} \left(\frac{1}{2}\right)(-2)\, du = \int_{0}^{4} -e^{u}\, du = -e^{u} \Big|_{0}^{4}$$

$$= -e^{-t/2} \Big|_{0}^{4} = -(e^{-4/2} - e^{-0/2}) = -(e^{-2} - 1) = 1 - e^{-2} \approx .86$$

Thus, if we were to randomly select an unemployed worker, there would be a .86 probability that the worker had been unemployed for 4 months or less. The second question asks for the probability that a randomly selected worker had been unemployed for 12 months or more, which is $\int_{12}^{\infty} \frac{1}{2} e^{-t/2}\, dt$. The calculation of integrals relating to regions of infinite extent (sometimes called *improper integrals*) is beyond the scope of this monograph, but we can make use of the second condition for probability density functions to find this area. Begin by calculating the area under the curve between 0 and 12. Picking up the integration from the relevant place from above,

$$-e^{-t/2} \Big|_{0}^{12} = -(e^{-12/2} - e^{-0/2}) = -(e^{-6} - 1) = 1 - e^{-6} \approx .998$$

The second condition for probability functions requires $\int_{-\infty}^{+\infty} f(t)\, dt = 1$. This condition is not met by $f(t) = \frac{1}{2} e^{-t/2}$ unless we restrict the domain to

$t > 0$. Because t in this problem represents time, and we do not consider negative time, we can rewrite this integral as $\int_{0}^{+\infty} \frac{1}{2} e^{-t/2} \, dt = 1$, which satisfies the condition. From rule 6 we also know that

$$\int_{0}^{12} \frac{1}{2} e^{-t/2} dt + \int_{12}^{+\infty} \frac{1}{2} e^{-t/2} dt = \int_{0}^{+\infty} \frac{1}{2} e^{-t/2} dt$$

We know the numerical values for two of the integrals in the above equation, so we need only substitute them into the equation to find the numerical value for the third:

$$.998 + \int_{12}^{+\infty} \frac{1}{2} e^{-t/2} dt \approx 1 \Rightarrow \int_{12}^{+\infty} \frac{1}{2} e^{-t/2} dt \approx 1 - .998 \approx .002$$

Thus, according to the formula, there is only about a .002 probability that a randomly selected worker has been unemployed 12 months or more.

Problem 5.2.4. Suppose that the revenues generated from a manufacturing machine t years from now are $f(t) = 3500 - 15t^2$ dollars per year, and the costs of running it t years from now are given by $g(t) = 1000 + 10t^2$. For how many years will use of the machine be profitable? What will the machine's net earnings be during the time it is profitable?

To answer the first question, we need to set the two functions equal to each other and solve

$$1000 + 10t^2 = 3500 - 15t^2 \Rightarrow 25t^2 = 2500 \Rightarrow t^2 = 100 \Rightarrow t = 10$$

Thus, the machine will remain profitable for 10 years. The second question is asking us to calculate the area *between* the two curves (see Figure 5.1). This is just an extension of what we have done previously with definite integrals. We know that the integral of a function over a specified interval is the area of the curve for that interval. What we *really* mean by "area under the curve" is the area *under* the curve and *above* the x-axis. (If it is not above the x-axis, then we consider it to be negative area as in Problem 5.2.1c.) Thus, we can think of a definite

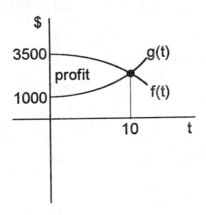

Figure 5.1. Graph for Problem 5.2.4

integral as $\int_a^b f(x)\,dx - \int_a^b g(x)\,dx$, where $g(x) = 0$ (i.e., the x-axis). By

extension, $\int_a^b f(x)\,dx - \int_a^b g(x)\,dx$ yields the area between the two curves

(under $f(x)$ and above $g(x)$), provided both functions are continuous over the specified interval. Returning to the present problem, we can represent the second question as

$$\int_0^{10} [f(t) - g(t)]\,dt = \int_0^{10} [(3500 - 15t^2) - (1000 + 10t^2)]\,dt = \int_0^{10} (2500 - 25t^2)\,dt$$

$$= 2500t - \frac{25}{3}t^3 \Big|_0^{10} = 2500(10) - \frac{25}{3}(10)^3 - 0 = 25,000 - \frac{25,000}{3} = \frac{50,000}{3}$$

$$= 16,666.67$$

Thus, over the 10-year period the machine will generate \$16,666.67 in profits.

5.3 Homework Problems

1. Use the rules for integration to evaluate the following indefinite integrals.

a. $\int 4\,dx$

f. $\int (x^{4/3} + x^{-2/3})\,dx$

b. $\int 8x\,dx$

g. $\int 2t^2(2t-1)^2\,dt$

c. $\int u^7\,du$

h. $\int \dfrac{z^3+3}{3z^5+9z^2}\,dz$

d. $\int 6z^{-3}\,dz$

i. $\int 7e^x\,dx$

e. $\int (6x^2 + 2x - 9)\,dx$

j. $\int (8x^{-5} + 4e^x)\,dx$

2. Use the chain rule to evaluate the following indefinite integrals.

a. $\int (7x-9)^3\,dx$

d. $\int t^2(6t^3-8)^4\,dt$

b. $\int e^{4t}\,dt$

e. $\int 9x^2 e^{3x^3}\,dx$

c. $\int 4z(2z^2-5)^4\,dz$

f. $\int \dfrac{(3x^{-2}+2)^3}{x^3}\,dx$

3. Use integration by parts to evaluate the following indefinite integrals.

a. $\int 2ze^{3z}\,dz$

c. $\int x(x+7)^{1/2}\,dx$

b. $\int (1+x)e^{-x}\,dx$

d. $\int t^2 e^{3t}\,dt$

4. Use the rules for integration to evaluate the following definite integrals.

a. $\int_0^3 (x+9)\,dx$

d. $\int_1^4 \dfrac{x+1}{x^{1/2}}\,dx$

b. $\int_{-1}^1 (2z^3-3)\,dz$

e. $\int_{\sqrt{5}}^{2\sqrt{3}} t(4+t^2)^{-3/2}\,dt$

c. $\int_{-3/4}^2 (-4t^2-5t-6)\,dt$

f. $\int_1^4 z^{1/2}\ln z\,dz$

5. An admissions officer at a small college determines that the number of applications for admission t years from now will be $1000 + \dfrac{1}{5}t$.

Rounded off to the nearest application, how many applications will the college's admissions office process in the next 5 years? How many applications above that number will the office process in the 5 years after that?

6. An agency estimates that sales of a new booklet will be $3000 - e^{x/5}$ copies x months from now. To the nearest month, how long before sales of the booklet drop to 0? How many booklets can the agency expect to sell in that time? (*Hint*: Integrate over the sales period.)

7. An investor estimates that t years from now two investments will generate income at rates of $200 + 5t$ and $50 + t^2$. How many years will it take for the second to generate the same amount of income as the first? In that time, how much more income will the first have generated? How many more complete years will it take before the total income of the second exceeds that of the first? (*Hint*: For the third part, find the value of t for which the difference in the two plans is 0.)

6. MATRIX ALGEBRA

6.1 Matrices

Recall the rules for matrices:

1. Two matrices $\mathbf{A} = [a_{ij}]$ and $\mathbf{B} = [b_{ij}]$ are equal iff $\mathbf{A}$ and $\mathbf{B}$ have the same order (i.e., the same number of rows and columns) and $[a_{ij}] = [b_{ij}]$ for all i and j.

2. For $\mathbf{A} = [a_{ij}]$ and $\mathbf{B} = [b_{ij}]$ with equal order, $m \times n$, the matrix $\mathbf{A} + \mathbf{B}$ is the matrix of order $m \times n$ such that $\mathbf{A} + \mathbf{B} = [a_{ij} + b_{ij}]$. In addition, the commutative law $(\mathbf{A} + \mathbf{B} = \mathbf{B} + \mathbf{A})$ and the associative law $[(\mathbf{A} + \mathbf{B}) + \mathbf{C} = \mathbf{A} + (\mathbf{B} + \mathbf{C})]$ both hold for matrix addition.

3. If $\mathbf{A} = [a_{ij}]$ and k is a scalar (constant), then the matrix $k\mathbf{A} = [ka_{ij}] = \mathbf{A}k$. If you multiply a matrix by a scalar, then *every* element of the matrix is multiplied by that scalar. Multiplication by a scalar is commutative.

4. Given rules 2 and 3, matrix subtraction can be defined as the addition of two matrices, one of which is multiplied by the scalar -1: $\mathbf{A} - \mathbf{B} = \mathbf{A} + (-1)\mathbf{B}$.

5. Matrix multiplication: If you have two matrices, $\mathbf{A} = [a_{ik}]$ of order $m \times p$ and $\mathbf{B} = [b_{kj}]$ of order $p \times n$, the matrix $\mathbf{AB} = \mathbf{C} = [c_{ij}]$ has

order $m \times n$, where $c_{ij} = \sum_{k=1}^{p} a_{ik}b_{kj}$. In calculating $\mathbf{AB}$, we say that $\mathbf{A}$ is *postmultiplied* by $\mathbf{B}$ or that $\mathbf{B}$ is *premultiplied* by $\mathbf{A}$.

6. Matrix multiplication is associative: $\mathbf{A(BC)} = \mathbf{(AB)C}$.

7. Matrix multiplication is distributive over matrix addition: If $\mathbf{AB} + \mathbf{AC}$ and $\mathbf{A(B + C)}$ both exist, then $\mathbf{AB} + \mathbf{AC} = \mathbf{A(B + C)}$.

8. A matrix is *symmetric* iff $a_{ij} = a_{ji}$ for all pairs (i, j). Both diagonal and identity matrices (below) are forms of symmetric matrices.

9. A *diagonal* matrix is a square matrix of the form

$$\mathbf{D} = \begin{bmatrix} d_{11} & 0 & 0 & \cdots & 0 \\ 0 & d_{22} & 0 & \cdots & 0 \\ 0 & 0 & \ddots & & \vdots \\ \vdots & \vdots & & \ddots & 0 \\ 0 & 0 & \cdots & 0 & d_{nn} \end{bmatrix}.$$

10. A special form of the diagonal matrix is the *identity* matrix: for example, $\mathbf{I}_3 = \begin{bmatrix} 1 & 0 & 0 \\ 0 & 1 & 0 \\ 0 & 0 & 1 \end{bmatrix}$. The identity matrix often is designated with only one subscript because the number of rows and columns must be equal.

11. The *transpose* of a matrix $\mathbf{A}$ is another matrix, designated by $\mathbf{A}^T$ or $\mathbf{A}'$, such that if $A_{m \times n} = [a_{ij}]$, then $A_{n \times m}^T = [a_{ji}]$.

Problem 6.1.1. Perform the following matrix operations on $\mathbf{A}$, $\mathbf{B}$, $\mathbf{C}$, and $\mathbf{D}$, where

$$\mathbf{A} = \begin{bmatrix} 3 & -2 & 1 \\ 0 & 4 & 2 \end{bmatrix} \quad \mathbf{B} = \begin{bmatrix} 1 & 2 & 1 \\ 5 & -1 & 3 \end{bmatrix} \quad \mathbf{C} = \begin{bmatrix} 2 & -3 \\ -1 & 1 \\ 1 & 4 \end{bmatrix} \quad \mathbf{D} = \begin{bmatrix} 1 & 2 \\ 3 & 4 \end{bmatrix}$$

a. $\mathbf{A + B}$: Both matrices are 2×3 so we can do the addition. From rule 1 in matrix addition, each element of the first matrix is added to the corresponding element of the second matrix. Thus,

$$\mathbf{A} + \mathbf{B} = \begin{bmatrix} 3 & -2 & 1 \\ 0 & 4 & 2 \end{bmatrix} + \begin{bmatrix} 1 & 2 & 1 \\ 5 & -1 & 3 \end{bmatrix} = \begin{bmatrix} 3+1 & -2+2 & 1+1 \\ 0+5 & 4-1 & 2+3 \end{bmatrix} = \begin{bmatrix} 4 & 0 & 2 \\ 5 & 3 & 5 \end{bmatrix}$$

b. **A + C**: The two matrices do not have the same dimensions so we cannot do the addition.

c. **B − A**: The matrices have the proper dimensions. From rule 4 we can do the subtraction as follows:

$$\mathbf{B} - \mathbf{A} = \begin{bmatrix} 1 & 2 & 1 \\ 5 & -1 & 3 \end{bmatrix} - \begin{bmatrix} 3 & -2 & 1 \\ 0 & 4 & 2 \end{bmatrix} = \begin{bmatrix} 1-3 & 2-(-2) & 1-1 \\ 5-0 & -1-4 & 3-2 \end{bmatrix} = \begin{bmatrix} -2 & 4 & 0 \\ 5 & -5 & 1 \end{bmatrix}$$

d. **3 × A**: From rule 3 we know that to multiply a matrix by a scalar (constant), each element of the matrix must be multiplied by the scalar. Thus,

$$3 \times \mathbf{A} = 3 \times \begin{bmatrix} 3 & -2 & 1 \\ 0 & 4 & 2 \end{bmatrix} = \begin{bmatrix} 3\times3 & 3\times-2 & 3\times1 \\ 3\times0 & 3\times4 & 3\times2 \end{bmatrix} = \begin{bmatrix} 9 & -6 & 3 \\ 0 & 12 & 6 \end{bmatrix}$$

e. **2A − 3B**: First multiply each matrix by the scalar:

$$2 \times \mathbf{A} = 2 \times \begin{bmatrix} 3 & -2 & 1 \\ 0 & 4 & 2 \end{bmatrix} = \begin{bmatrix} 2\times3 & 2\times-2 & 2\times1 \\ 2\times0 & 2\times4 & 2\times2 \end{bmatrix} = \begin{bmatrix} 6 & -4 & 2 \\ 0 & 8 & 4 \end{bmatrix}$$

$$3 \times \mathbf{B} = 3 \times \begin{bmatrix} 1 & 2 & 1 \\ 5 & -1 & 3 \end{bmatrix} = \begin{bmatrix} 3\times1 & 3\times2 & 3\times1 \\ 3\times5 & 3\times-1 & 3\times3 \end{bmatrix} = \begin{bmatrix} 3 & 6 & 3 \\ 15 & -3 & 9 \end{bmatrix}$$

Now perform the subtraction:

$$\begin{bmatrix} 6 & -4 & 2 \\ 0 & 8 & 4 \end{bmatrix} - \begin{bmatrix} 3 & 6 & 3 \\ 15 & -3 & 9 \end{bmatrix} = \begin{bmatrix} 6-3 & -4-6 & 2-3 \\ 0-15 & 8-(-3) & 4-9 \end{bmatrix} = \begin{bmatrix} 3 & -10 & -1 \\ -15 & 11 & -5 \end{bmatrix}$$

f. **AB**: Although these matrices have the same dimensions, they are not correct for matrix multiplication. From rule 5, to postmultiply **A** by **B**, the number of columns of **A** must equal the number of rows of **B**. Here, **A** has three columns and **B** has two rows, so they are said to be *nonconformable* and we cannot do the multiplication.

g. **DA**: These matrices are conformable (i.e., they have the correct dimensions), so we proceed as follows:

$$\mathbf{DA} = \begin{bmatrix} 1 & 2 \\ 3 & 4 \end{bmatrix}\begin{bmatrix} 3 & -2 & 1 \\ 0 & 4 & 2 \end{bmatrix} = \begin{bmatrix} (1 \times 3) + (2 \times 0) & (1 \times -2) + (2 \times 4) & (1 \times 1) + (2 \times 2) \\ (3 \times 3) + (4 \times 0) & (3 \times -2) + (4 \times 4) & (3 \times 1) + (4 \times 2) \end{bmatrix}$$

$$= \begin{bmatrix} 3+0 & -2+8 & 1+4 \\ 9+0 & -6+16 & 3+8 \end{bmatrix} = \begin{bmatrix} 3 & 6 & 5 \\ 9 & 10 & 11 \end{bmatrix}$$

Notice that the resulting matrix has the same number of rows as the first matrix and the same number of columns as the second matrix.

h. **CD**: These matrices are conformable, so do the multiplication as follows:

$$\mathbf{CD} = \begin{bmatrix} 2 & -3 \\ -1 & 1 \\ 1 & 4 \end{bmatrix}\begin{bmatrix} 1 & 2 \\ 3 & 4 \end{bmatrix} = \begin{bmatrix} (2 \times 1) + (-3 \times 3) & (2 \times 2) + (-3 \times 4) \\ (-1 \times 1) + (1 \times 3) & (-1 \times 2) + (1 \times 4) \\ (1 \times 1) + (4 \times 3) & (1 \times 2) + (4 \times 4) \end{bmatrix}$$

$$= \begin{bmatrix} 2 + (-9) & 4 + (-12) \\ -1 + 3 & -2 + 4 \\ 1 + 12 & 2 + 16 \end{bmatrix} = \begin{bmatrix} -7 & -8 \\ 2 & 2 \\ 13 & 18 \end{bmatrix}$$

6.2 Determinants

Problem 6.2.1. Find the determinants of the following matrices.

a. $\mathbf{A} = \begin{bmatrix} 1 & 2 \\ -1 & 3 \end{bmatrix}$: This is a square matrix, so we can calculate the determinant as

$$|\mathbf{A}| = \begin{vmatrix} 1 & 2 \\ -1 & 3 \end{vmatrix} = (1 \times 3) - (2 \times -1) = 3 + 2 = 5.$$

b. $\mathbf{B} = \begin{bmatrix} 3 & 0 \\ 7 & 1 \end{bmatrix} \Rightarrow |\mathbf{B}| = \begin{vmatrix} 3 & 0 \\ 7 & 1 \end{vmatrix} = (3 \times 1) - (0 \times 7) = 3 - 0 = 3$

c. $\mathbf{C} = \begin{bmatrix} 0 & 3 & -1 \\ 1 & 1 & 2 \end{bmatrix}$: Because this matrix is not square, it does not have a determinant.

d. $\mathbf{D} = \begin{bmatrix} 2 & 6 \\ 1 & 3 \end{bmatrix} \Rightarrow |\mathbf{D}| = \begin{vmatrix} 2 & 6 \\ 1 & 3 \end{vmatrix} = (2 \times 3) - (6 \times 1) = 6 - 6 = 0$: Determinants can be 0, but some later calculations will require matrices with nonzero determinants.

Problem 6.2.2. Find the determinants of the following 3×3 matrices using one of the two common methods.

a. $\mathbf{A} = \begin{bmatrix} 1 & 1 & 5 \\ 3 & -2 & -1 \\ 2 & 4 & 2 \end{bmatrix}$: Use the butterfly method to find the determinant of $\mathbf{A}$:

$$|\mathbf{A}| = \begin{vmatrix} 1 & 1 & 5 \\ 3 & -2 & -1 \\ 2 & 4 & 2 \end{vmatrix}$$

$$= (1 \times -2 \times 2) + (1 \times -1 \times 2) + (5 \times 4 \times 3) - (2 \times -2 \times 5) - (4 \times -1 \times 1) - (2 \times 1 \times 3)$$

$$= (-4) + (-2) + 60 - (-20) - (-4) - 6 = 72$$

b. $\mathbf{B} = \begin{bmatrix} 0 & -1 & 2 \\ -2 & 3 & 4 \\ 1 & 2 & 1 \end{bmatrix}$: Use the butterfly method to find the determinant of $\mathbf{B}$:

$$|\mathbf{B}| = \begin{vmatrix} 0 & -1 & 2 \\ -2 & 3 & 4 \\ 1 & 2 & 1 \end{vmatrix}$$

$$= (0 \times 3 \times 1) + (-1 \times 4 \times 1) + (2 \times 2 \times -2) - (1 \times 3 \times 2) - (2 \times 4 \times 0) - (1 \times -1 \times -2)$$

$$= 0 + (-4) + (-8) - 6 - 0 - 2 = -20$$

c. $\mathbf{C} = \begin{bmatrix} 2 & 1 & 0 \\ -1 & 1 & 3 \\ 0 & 2 & 2 \end{bmatrix}$: Use the method of rewriting the first two columns to find the determinant of $\mathbf{C}$:

$$|\mathbf{C}| = \begin{vmatrix} 2 & 1 & 0 \\ -1 & 1 & 3 \\ 0 & 2 & 2 \end{vmatrix} \begin{matrix} 2 & 1 \\ -1 & 1 \\ 0 & 2 \end{matrix}$$

$$= (2 \times 1 \times 2) + (1 \times 3 \times 0) + (0 \times -1 \times 2) - (0 \times 1 \times 0) - (2 \times 3 \times 2) - (2 \times -1 \times 1)$$

$$= 4 + 0 + 0 - 0 - 12 - (-2) = -6$$

d. $\mathbf{D} = \begin{bmatrix} -1 & 2 & 1 \\ -2 & 4 & 0 \\ 1 & 3 & -2 \end{bmatrix}$: Use the method of rewriting the first two columns to find the determinant of $\mathbf{D}$:

$$|\mathbf{D}| = \begin{vmatrix} -1 & 2 & 1 \\ -2 & 4 & 0 \\ 1 & 3 & -2 \end{vmatrix}\begin{matrix} -1 & 2 \\ -2 & 4 \\ 1 & 3 \end{matrix}$$

$$= (-1 \times 4 \times -2) + (2 \times 0 \times 1) + (1 \times -2 \times 3) - (1 \times 4 \times 1) - (3 \times 0 \times -1) - (-2 \times -2 \times 2)$$

$$= 8 + 0 + (-6) - 4 - 0 - 8 = -10$$

Problem 6.2.3. Use row or column expansion to find the determinants of the following matrices.

a. $\mathbf{A} = \begin{bmatrix} 1 & -3 & 2 \\ -3 & 3 & -1 \\ 2 & -1 & 0 \end{bmatrix}$: Expand by the first row without first manipulating the rows and columns:

$$|\mathbf{A}| = \begin{vmatrix} 1 & -3 & 2 \\ -3 & 3 & -1 \\ 2 & -1 & 0 \end{vmatrix} = (-1)^{1+1} \times 1 \times \begin{vmatrix} 3 & -1 \\ -1 & 0 \end{vmatrix} + (-1)^{1+2} \times (-3) \times \begin{vmatrix} -3 & -1 \\ 2 & 0 \end{vmatrix} +$$

$$(-1)^{1+3} \times 2 \times \begin{vmatrix} -3 & 3 \\ 2 & -1 \end{vmatrix}$$

$$= 1 \times \begin{vmatrix} 3 & -1 \\ -1 & 0 \end{vmatrix} + 3 \times \begin{vmatrix} -3 & -1 \\ 2 & 0 \end{vmatrix} + 2 \times \begin{vmatrix} -3 & 3 \\ 2 & -1 \end{vmatrix}$$

$$= 1 \times (-1) + 3 \times (2) + 2 \times (-3) = -1 + 6 - 6 = -1$$

Notice that after the expansion we are left with three 2×2 matrices. Using row or column expansion on a 4×4 matrix would result in four 3×3 matrices, which also would need to be evaluated.

b. Find the determinant of the matrix in Problem 6.2.3a using row expansion after manipulating the rows and columns. Remember that the goal of the manipulation is to get a row or column with a 1 in one position and 0s in the rest. Begin by adding the second column to the first:

$$|\mathbf{A}| = \begin{vmatrix} 1 & -3 & 2 \\ -3 & 3 & -1 \\ 2 & -1 & 0 \end{vmatrix} \xrightarrow[C_2 + C_1]{} \begin{vmatrix} -2 & -3 & 2 \\ 0 & 3 & -1 \\ 1 & -1 & 0 \end{vmatrix}$$

Notice the "directions" under the arrow indicating the operation to be performed on the rows ($\mathbf{R}_i$) or columns ($\mathbf{C}_i$). The second listed row or column is where the new row or column will be located. Now add the first column to the second:

$$\begin{vmatrix} -2 & -3 & 2 \\ 0 & 3 & -1 \\ 1 & -1 & 0 \end{vmatrix} \xrightarrow[C_1 + C_2]{} \begin{vmatrix} -2 & -5 & 2 \\ 0 & 3 & -1 \\ 1 & 0 & 0 \end{vmatrix}$$

Now expand by the third row:

$$\begin{vmatrix} -2 & -5 & 2 \\ 0 & 3 & -1 \\ 1 & 0 & 0 \end{vmatrix} = (-1)^{3+1} \times 1 \times \begin{vmatrix} -5 & 2 \\ 3 & -1 \end{vmatrix} = \begin{vmatrix} -5 & 2 \\ 3 & -1 \end{vmatrix} = 5 - 6 = -1$$

c. $\mathbf{C} = \begin{bmatrix} 3 & 1 & 2 & 4 \\ 2 & 0 & 5 & 1 \\ 1 & -1 & -2 & 6 \\ -2 & 3 & 2 & 3 \end{bmatrix}$: Manipulate the rows and columns to expand by the

second column. Let us do two steps at once to save time. Add the first row to the third row and add minus 3 times the first row to the fourth row:

$$|\mathbf{C}| = \begin{vmatrix} 3 & 1 & 2 & 4 \\ 2 & 0 & 5 & 1 \\ 1 & -1 & -2 & 6 \\ -2 & 3 & 2 & 3 \end{vmatrix} \xrightarrow[\substack{R_1 + R_3 \\ -3R_1 + R_4}]{} \begin{vmatrix} 3 & 1 & 2 & 4 \\ 2 & 0 & 5 & 1 \\ 4 & 0 & 0 & 10 \\ -11 & 0 & -4 & -9 \end{vmatrix}$$

Now expand by the second column:

$$|\mathbf{C}| = \begin{vmatrix} 3 & 1 & 2 & 4 \\ 2 & 0 & 5 & 1 \\ 4 & 0 & 0 & 10 \\ -11 & 0 & -4 & -9 \end{vmatrix} = (-1)^{1+2} \times 1 \times \begin{vmatrix} 2 & 5 & 1 \\ 4 & 0 & 10 \\ -11 & -4 & -9 \end{vmatrix} = (-1) \times \begin{vmatrix} 2 & 5 & 1 \\ 4 & 0 & 10 \\ -11 & -4 & -9 \end{vmatrix}$$

We are left with a 3×3 determinant, so manipulate the rows to expand by the third column. Add minus 10 times the first row to the second row and add 9 times the first row to the third row:

$$(-1) \times \begin{vmatrix} 2 & 5 & 1 \\ 4 & 0 & 10 \\ -11 & -4 & -9 \end{vmatrix} \xrightarrow[\substack{-10R_1 + R_2 \\ 9R_1 + R_3}]{} (-1) \begin{vmatrix} 2 & 5 & 1 \\ -16 & -50 & 0 \\ 7 & 41 & 0 \end{vmatrix}$$

Now expand by the third column:

$$(-1) \begin{vmatrix} 2 & 5 & 1 \\ -16 & -50 & 0 \\ 7 & 41 & 0 \end{vmatrix} = (-1)(-1)^{1+3} \times 1 \times \begin{vmatrix} -16 & -50 \\ 7 & 41 \end{vmatrix}$$

$$= (-1) \times [(-16 \times 41) - (7 \times -50)] = (-1) \times (-656 + 350) = 306$$

(*Note*: It may sound a bit odd to say "add minus," but it allows the directions to be clearer about the placement of the new row or column.)

6.3 Inverses of Matrices

Problem 6.3.1. Find the inverses of the following matrices.

a. $A = \begin{bmatrix} 3 & 1 \\ 6 & 2 \end{bmatrix}$: A matrix must be *nonsingular* (i.e., $|A| \neq 0$) for its inverse to exist. Here, $|A| = 0$, so A does not have an inverse.

b. $B = \begin{bmatrix} 3 & 2 \\ 5 & 4 \end{bmatrix}$: Begin by determining that $|B| = 2$. Next form the cofactor matrix of B:

$$\text{cof } B = \begin{bmatrix} (-1)^{1+1} \times 4 & (-1)^{1+2} \times 5 \\ (-1)^{2+1} \times 2 & (-1)^{2+2} \times 3 \end{bmatrix} = \begin{bmatrix} 4 & -5 \\ -2 & 3 \end{bmatrix}$$

Now form the adjoint of B:

$$\text{adj } B = (\text{cof } B)^T = \begin{bmatrix} 4 & -5 \\ -2 & 3 \end{bmatrix}^T = \begin{bmatrix} 4 & -2 \\ -5 & 3 \end{bmatrix}$$

Complete the calculation by multiplying adj B by 1 over the determinant of B:

$$B^{-1} = \left(\frac{1}{|B|}\right)(\text{adj } B) = \left(\frac{1}{2}\right)\begin{bmatrix} 4 & -2 \\ -5 & 3 \end{bmatrix} = \begin{bmatrix} \frac{4}{2} & \frac{-2}{2} \\ \frac{-5}{2} & \frac{3}{2} \end{bmatrix} = \begin{bmatrix} 2 & -1 \\ \frac{-5}{2} & \frac{3}{2} \end{bmatrix}$$

Verify for yourself that $BB^{-1} = I$

c. $\mathbf{C} = \begin{bmatrix} 1 & 2 & 3 \\ 2 & 4 & 1 \\ 1 & 3 & 0 \end{bmatrix}$: First determine that $|\mathbf{C}| = 5$. Now form the cofactor matrix of $\mathbf{C}$:

$$\text{cof } \mathbf{C} = \begin{bmatrix} (-1)^{1+1}((4 \times 0)-(1 \times 3)) & (-1)^{1+2}((2 \times 0)-(1 \times 1)) & (-1)^{1+3}((2 \times 3)-(1 \times 4)) \\ (-1)^{2+1}((2 \times 0)-(3 \times 3)) & (-1)^{2+2}((1 \times 0)-(1 \times 3)) & (-1)^{2+3}((1 \times 3)-(1 \times 2)) \\ (-1)^{3+1}((2 \times 1)-(4 \times 3)) & (-1)^{3+2}((1 \times 1)-(2 \times 3)) & (-1)^{3+3}((1 \times 4)-(2 \times 2)) \end{bmatrix}$$

$$= \begin{bmatrix} -3 & 1 & 2 \\ 9 & -3 & -1 \\ -10 & 5 & 0 \end{bmatrix}$$

Now find the adjoint of $\mathbf{C}$ and divide by the determinant of $\mathbf{C}$:

$$\mathbf{C}^{-1} = \frac{\text{adj } \mathbf{C}}{|\mathbf{C}|} = \frac{(\text{cof } \mathbf{C})^{\mathrm{T}}}{|\mathbf{C}|} = \left(\frac{1}{5}\right)\begin{bmatrix} -3 & 1 & 2 \\ 9 & -3 & -1 \\ -10 & 5 & 0 \end{bmatrix}^{\mathrm{T}} = \left(\frac{1}{5}\right)\begin{bmatrix} -3 & 9 & -10 \\ 1 & -3 & 5 \\ 2 & -1 & 0 \end{bmatrix} = \begin{bmatrix} \frac{-3}{5} & \frac{9}{5} & -2 \\ \frac{1}{5} & \frac{-3}{5} & 1 \\ \frac{2}{5} & \frac{-1}{5} & 0 \end{bmatrix}$$

Verify for yourself that $\mathbf{CC}^{-1} = \mathbf{I}$.

d. $\mathbf{D} = \begin{bmatrix} 2 & 3 & 4 \\ 4 & 3 & 1 \\ -1 & 2 & 5 \end{bmatrix}$: Determine that $|\mathbf{D}| = 7$, then form the cofactor matrix of $\mathbf{D}$:

$$\text{cof } \mathbf{D} = \begin{bmatrix} (-1)^{1+1}((3 \times 5)-(2 \times 1)) & (-1)^{1+2}((4 \times 5)-(-1 \times 1)) & (-1)^{1+3}((4 \times 2)-(-1 \times 3)) \\ (-1)^{2+1}((3 \times 5)-(2 \times 4)) & (-1)^{2+2}((2 \times 5)-(-1 \times 4)) & (-1)^{2+3}((2 \times 2)-(-1 \times 3)) \\ (-1)^{3+1}((3 \times 1)-(3 \times 4)) & (-1)^{3+2}((2 \times 1)-(4 \times 4)) & (-1)^{3+3}((2 \times 3)-(4 \times 3)) \end{bmatrix}$$

$$= \begin{bmatrix} 13 & -21 & 11 \\ -7 & 14 & -7 \\ -9 & 14 & -6 \end{bmatrix}$$

Now find the adjoint of $\mathbf{D}$ and divide by the determinant of $\mathbf{D}$:

$$\mathbf{D}^{-1} = \frac{\text{adj } \mathbf{D}}{|\mathbf{D}|} = \frac{(\text{cof } \mathbf{D})^{\mathrm{T}}}{|\mathbf{D}|} = \left(\frac{1}{7}\right)\begin{bmatrix} 13 & -21 & 11 \\ -7 & 14 & -7 \\ -9 & 14 & -6 \end{bmatrix}^{\mathrm{T}} = \left(\frac{1}{7}\right)\begin{bmatrix} 13 & -7 & -9 \\ -21 & 14 & 14 \\ 11 & -7 & -6 \end{bmatrix} = \begin{bmatrix} \dfrac{13}{7} & -1 & \dfrac{-9}{7} \\ -3 & 2 & 2 \\ \dfrac{11}{7} & -1 & \dfrac{-6}{7} \end{bmatrix}$$

Verify for yourself that $\mathbf{DD}^{-1} = \mathbf{I}$.

Row Operations. Although there will be times when we need to know the adjoint and cofactor matrices, this method can be very tedious, particularly for matrices larger than 3×3. An alternate method is to use *row operations*. The process of manipulating the rows of a matrix to get 1s and 0s in convenient places can be extended to finding inverses. If a sequence of row operations reduces $\mathbf{A}$ to $\mathbf{I}$, then the same sequence of row operations reduces the partitioned matrix $[\mathbf{A} \vdots \mathbf{I}]$ to $[\mathbf{I} \vdots \mathbf{A}^{-1}]$.

Problem 6.3.2. Use row operations to find the inverses of the following matrices.

a. $\mathbf{A} = \begin{bmatrix} 1 & -1 & 2 & 3 \\ 2 & -1 & 0 & 2 \\ 4 & 1 & -11 & -1 \\ 1 & 2 & 3 & 83 \end{bmatrix}$: Begin by constructing $[\mathbf{A} \vdots \mathbf{I}]$ as follows:

$$[\mathbf{A} \vdots \mathbf{I}] = \begin{bmatrix} 1 & -1 & 2 & 3 & \vdots & 1 & 0 & 0 & 0 \\ 2 & -1 & 0 & 2 & \vdots & 0 & 1 & 0 & 0 \\ 4 & 1 & -11 & -1 & \vdots & 0 & 0 & 1 & 0 \\ 1 & 2 & 3 & 83 & \vdots & 0 & 0 & 0 & 1 \end{bmatrix}$$

Now use row operations to reduce the left side of the matrix to $\mathbf{I}_4$. Start by adding $-2\mathbf{R}_1$ to $\mathbf{R}_2$, $-4\mathbf{R}_1$ to $\mathbf{R}_3$, and $-\mathbf{R}_1$ to $\mathbf{R}_4$:

$$\begin{array}{c} \xrightarrow{} \\ -2\mathbf{R}_1 + \mathbf{R}_2 \\ -4\mathbf{R}_1 + \mathbf{R}_3 \\ -\mathbf{R}_1 + \mathbf{R}_4 \end{array} \begin{bmatrix} 1 & -1 & 2 & 3 & \vdots & 1 & 0 & 0 & 0 \\ 0 & 1 & -4 & -4 & \vdots & -2 & 1 & 0 & 0 \\ 0 & 5 & -19 & -13 & \vdots & -4 & 0 & 1 & 0 \\ 0 & 3 & 1 & 80 & \vdots & -1 & 0 & 0 & 1 \end{bmatrix}$$

Notice that the operations were selected to create 0s in the rest of the first column. Next work with the second row to obtain 0s in the first, third, and fourth positions in the second column. Then continue the process using the third and fourth rows:

$$\xrightarrow[\substack{R_2 + R_1 \\ -5R_2 + R_3 \\ -3R_2 + R_4}]{}\left[\begin{array}{cccc|cccc} 1 & 0 & -2 & -1 & -1 & 1 & 0 & 0 \\ 0 & 1 & -4 & -4 & -2 & 1 & 0 & 0 \\ 0 & 0 & 1 & 7 & 6 & -5 & 1 & 0 \\ 0 & 0 & 13 & 92 & 5 & -3 & 0 & 1 \end{array}\right]$$

$$\xrightarrow[\substack{2R_3 + R_1 \\ 4R_3 + R_2 \\ -13R_3 + R_4}]{}\left[\begin{array}{cccc|cccc} 1 & 0 & 0 & 13 & 11 & -9 & 2 & 0 \\ 0 & 1 & 0 & 24 & 22 & -19 & 4 & 0 \\ 0 & 0 & 1 & 7 & 6 & -5 & 1 & 0 \\ 0 & 0 & 0 & 1 & -73 & 62 & -13 & 1 \end{array}\right]$$

$$\xrightarrow[\substack{-7R_4 + R_3 \\ -24R_4 + R_2 \\ -13R_4 + R_1}]{}\left[\begin{array}{cccc|cccc} 1 & 0 & 0 & 0 & 960 & -815 & 171 & -13 \\ 0 & 1 & 0 & 0 & 1774 & -1507 & 316 & -24 \\ 0 & 0 & 1 & 0 & 517 & -439 & 92 & -7 \\ 0 & 0 & 0 & 1 & -73 & 62 & -13 & 1 \end{array}\right] = [\mathbf{I} \vdots \mathbf{A}^{-1}].$$

Obviously, this problem was constructed to have a 1 appear in the proper location for each set of operations and not to require the use of fractions.

b. $\mathbf{B} = \begin{bmatrix} 7 & -8 & -5 \\ -3 & 6 & 3 \\ -1 & 2 & -1 \end{bmatrix}$: As before, first construct $[\mathbf{B} \vdots \mathbf{I}]$. We do not have a 1 in

the upper-left corner, so begin by multiplying the third row by -1, then exchange the third and first rows ($R_3 \leftrightarrow R_1$) and continue with the necessary row operations:

$$[\mathbf{B} \vdots \mathbf{I}] = \begin{bmatrix} 7 & -8 & -5 & 1 & 0 & 0 \\ -3 & 6 & 3 & 0 & 1 & 0 \\ -1 & 2 & -1 & 0 & 0 & 1 \end{bmatrix} \xrightarrow[\substack{-R_3 \\ R_3 \leftrightarrow R_1}]{} \begin{bmatrix} 1 & -2 & 1 & 0 & 0 & -1 \\ -3 & 6 & 3 & 0 & 1 & 0 \\ 7 & -8 & -5 & 1 & 0 & 0 \end{bmatrix}$$

$$\xrightarrow[\substack{3R_1 + R_2 \\ -7R_1 + R_3}]{} \begin{bmatrix} 1 & -2 & 1 & 0 & 0 & -1 \\ 0 & 0 & 6 & 0 & 1 & -3 \\ 0 & 6 & -12 & 1 & 0 & 7 \end{bmatrix}$$

This time there are no 1s in the second column. This means we will have fractions in the inverse. There are two ways of proceeding: (a) continue as usual and work with the fractions as they appear, or (b) reduce $\mathbf{B}$ to $6\mathbf{I}$

rather than **I** (i.e., 6 times the identity matrix), which will allow us to avoid fractions temporarily. Let us use the second method and begin by exchanging the third and second rows. Next multiply the *first* row by 6, then work with the 6 in the second position of the second row to obtain 0s in the other positions of the second column:

$$\xrightarrow[\substack{R_3 \leftrightarrow R_2 \\ 6R_1}]{} \left[\begin{array}{ccc|ccc} 6 & -12 & 6 & 0 & 0 & -6 \\ 0 & 6 & -12 & 1 & 0 & 7 \\ 0 & 0 & 6 & 0 & 1 & -3 \end{array}\right] \xrightarrow{2R_2 + R_1} \left[\begin{array}{ccc|ccc} 6 & 0 & -18 & 2 & 0 & 8 \\ 0 & 6 & -12 & 1 & 0 & 7 \\ 0 & 0 & 6 & 0 & 1 & -3 \end{array}\right]$$

We have a 6 in the third position of the third row, so complete the operations to obtain 0s in the rest of the third column, then multiply all three rows by $\frac{1}{6}$:

$$\xrightarrow[\substack{3R_3 + R_1 \\ 2R_3 + R_2}]{} \left[\begin{array}{ccc|ccc} 6 & 0 & 0 & 2 & 3 & -1 \\ 0 & 6 & 0 & 1 & 2 & 1 \\ 0 & 0 & 6 & 0 & 1 & -3 \end{array}\right] \xrightarrow[\substack{\frac{1}{6}R_1 \\ \frac{1}{6}R_2 \\ \frac{1}{6}R_3}]{} \left[\begin{array}{ccc|ccc} 1 & 0 & 0 & \frac{2}{6} & \frac{3}{6} & \frac{-1}{6} \\[4pt] 0 & 1 & 0 & \frac{1}{6} & \frac{2}{6} & \frac{1}{6} \\[4pt] 0 & 0 & 1 & 0 & \frac{1}{6} & \frac{-3}{6} \end{array}\right] = [\mathbf{I} : \mathbf{B}^{-1}].$$

Verify for yourself that $\mathbf{BB}^{-1} = \mathbf{I}$. (*Note:* Exchanging rows while using row operations to find an inverse does not change the result. Exchanging the rows of a *determinant*, however, changes its sign.)

6.4 Cramer's Rule

Problem 6.4.1. If $x_1 + x_2 + 2x_3 = 4$, $-x_1 + 2x_2 + 3x_3 = 1$, and $x_1 + x_2 - x_3 = 5$, find x, y, and z using Cramer's rule. Begin by putting the three equations in matrix form:

$$\begin{array}{l} x_1 + x_2 + 2x_3 = 4 \\[4pt] -x_1 + 2x_2 + 3x_3 = 1 \Rightarrow \mathbf{AX} = \mathbf{B} \Rightarrow \left[\begin{array}{ccc} 1 & 1 & 2 \\ -1 & 2 & 3 \\ 1 & 1 & -1 \end{array}\right]\left[\begin{array}{c} x_1 \\ x_2 \\ x_3 \end{array}\right] = \left[\begin{array}{c} 4 \\ 1 \\ 5 \end{array}\right] \\[4pt] x_1 + x_2 - x_3 = 5 \end{array}$$

Now solve for x_1 by substituting **B** for the first column of **A**, taking the determinant of the new matrix $\mathbf{A}_{(1 \leftrightarrow B)}$, and dividing by the determinant of **A**:

$$x_1 = \frac{|\mathbf{A}_{(1 \leftrightarrow B)}|}{|\mathbf{A}|} = \frac{\begin{vmatrix} 4 & 1 & 2 \\ 1 & 2 & 3 \\ 5 & 1 & -1 \end{vmatrix}}{\begin{vmatrix} 1 & 1 & 2 \\ -1 & 2 & 3 \\ 1 & 1 & -1 \end{vmatrix}} = \frac{-22}{-9} = \frac{22}{9}$$

Next calculate x_2 and x_3 by exchanging **B** for the second and third columns of **A**:

$$x_2 = \frac{|\mathbf{A}_{(2 \leftrightarrow B)}|}{|\mathbf{A}|} = \frac{\begin{vmatrix} 1 & 4 & 2 \\ -1 & 1 & 3 \\ 1 & 5 & -1 \end{vmatrix}}{-9} = \frac{-20}{-9} = \frac{20}{9},$$

$$x_3 = \frac{|\mathbf{A}_{(3 \leftrightarrow B)}|}{|\mathbf{A}|} = \frac{\begin{vmatrix} 1 & 1 & 4 \\ -1 & 2 & 1 \\ 1 & 1 & 5 \end{vmatrix}}{-9} = \frac{3}{-9} = \frac{-3}{9}$$

Verify these answers by inserting them into the original equations:

$$\frac{22}{9} + \frac{20}{9} + 2\left(\frac{-3}{9}\right) = \frac{42-6}{9} = \frac{36}{9} = 4$$

$$\frac{-22}{9} + 2\left(\frac{20}{9}\right) + 3\left(\frac{-3}{9}\right) = \frac{-22+40-9}{9} = \frac{9}{9} = 1$$

$$\frac{22}{9} + \frac{20}{9} - \frac{-3}{9} = \frac{22+20+3}{9} = \frac{45}{9} = 5$$

Problem 6.4.2. A political scientist is planning a small survey of small, medium, and large businesses in three cities. The budget for the surveys in each city varies based on the costs of obtaining information from each size of business. The budget for the first city is $150, with costs for small, medium, and large businesses of $10, $17, and $7, respectively. The budget for the second city is \$200, with costs of \$14, \$20, and \$10. The budget for the third city is \$100, with costs of \$9, \$9, and \$4. Using Cramer's rule, determine how many businesses of each size the political scientist can survey if the same number of each size must be included for all three cities?

Begin by constructing equations and putting them into matrix form. Let x_1 represent small businesses, x_2 represent medium businesses, and x_3 represent large businesses, then:

$$10x_1 + 17x_2 + 7x_3 = 150$$

$$14x_1 + 20x_2 + 10x_3 = 200 \Rightarrow \mathbf{AX} = \mathbf{B} \Rightarrow \begin{bmatrix} 10 & 17 & 7 \\ 14 & 20 & 10 \\ 9 & 9 & 4 \end{bmatrix} \begin{bmatrix} x_1 \\ x_2 \\ x_3 \end{bmatrix} = \begin{bmatrix} 150 \\ 200 \\ 100 \end{bmatrix}$$

$$9x_1 + 9x_2 + 4x_3 = 100$$

Now solve for x_1 by substituting $\mathbf{B}$ for the first column of $\mathbf{A}$, taking the determinant of the new matrix $\mathbf{A}_{(1 \leftrightarrow B)}$, and dividing by the determinant of $\mathbf{A}$:

$$x_1 = \frac{|\mathbf{A}_{(1 \leftrightarrow B)}|}{|\mathbf{A}|} = \frac{\begin{vmatrix} 150 & 17 & 7 \\ 200 & 20 & 10 \\ 100 & 9 & 4 \end{vmatrix}}{\begin{vmatrix} 10 & 17 & 7 \\ 14 & 20 & 10 \\ 7 & 10 & 5 \end{vmatrix}} = \frac{500}{100} = 5$$

Next calculate x_2 and x_3 by exchanging $\mathbf{B}$ for the second and third columns of $\mathbf{A}$:

$$x_2 = \frac{|\mathbf{A}_{(2 \leftrightarrow B)}|}{|\mathbf{A}|} = \frac{\begin{vmatrix} 10 & 150 & 7 \\ 14 & 200 & 10 \\ 9 & 100 & 4 \end{vmatrix}}{100} = \frac{300}{100} = 3$$

$$x_3 = \frac{|\mathbf{A}_{(3 \leftrightarrow B)}|}{|\mathbf{A}|} = \frac{\begin{vmatrix} 10 & 17 & 150 \\ 14 & 20 & 200 \\ 9 & 9 & 100 \end{vmatrix}}{100} = \frac{700}{100} = 7$$

Verify these answers by inserting them into the original equations:

$$10x_1 + 17x_2 + 7x_3 = 10(5) + 17(3) + 7(7) = 50 + 51 + 49 = 150$$
$$14x_1 + 20x_2 + 10x_3 = 14(5) + 20(3) + 10(7) = 70 + 60 + 70 = 200$$
$$9x_1 + 9x_2 + 4x_3 = 9(5) + 9(3) + 4(7) = 45 + 27 + 28 = 100$$

Thus, the political scientist must choose 5 small-sized, 3 medium-sized, and 7 large-sized businesses in each city.

6.5 Eigenvalues and Eigenvectors

Problem 6.5.1. Find the eigenvalues and eigenvectors of $A = \begin{bmatrix} 2 & 1 \\ 3 & 4 \end{bmatrix}$.

Begin by constructing $|A - \lambda I|$: $\begin{vmatrix} 2 & 1 \\ 3 & 4 \end{vmatrix} - \lambda \begin{vmatrix} 1 & 0 \\ 0 & 1 \end{vmatrix} = \begin{vmatrix} 2 & 1 \\ 3 & 4 \end{vmatrix} - \begin{vmatrix} \lambda & 0 \\ 0 & \lambda \end{vmatrix} = $
$\begin{vmatrix} 2 - \lambda & 1 \\ 3 & 4 - \lambda \end{vmatrix}$. Now solve the equation $|A - \lambda I| = 0$:

$$|A - \lambda I| = 0 = \begin{vmatrix} 2 - \lambda & 1 \\ 3 & 4 - \lambda \end{vmatrix} = (2 - \lambda)(4 - \lambda) - 3 = (8 - 6\lambda + \lambda^2) - 3$$

$$= \lambda^2 - 6\lambda + 5 = (\lambda - 5)(\lambda - 1) = 0$$

$$\therefore \lambda = 1, 5$$

Use the equation $(A - \lambda I)X = 0$ to determine the eigenvectors. For $\lambda = 1$:

$$\begin{bmatrix} 2 - 1 & 1 \\ 3 & 4 - 1 \end{bmatrix} \begin{bmatrix} x_1 \\ x_2 \end{bmatrix} = \begin{bmatrix} 0 \\ 0 \end{bmatrix} \Rightarrow \begin{bmatrix} 1 & 1 \\ 3 & 3 \end{bmatrix} \begin{bmatrix} x_1 \\ x_2 \end{bmatrix} = \begin{bmatrix} 0 \\ 0 \end{bmatrix} \Rightarrow \begin{bmatrix} x_1 + x_2 \\ 3x_1 + 3x_2 \end{bmatrix} = \begin{bmatrix} 0 \\ 0 \end{bmatrix}$$

Using the first equation, we know $x_1 + x_2 = 0 \Rightarrow x_1 = -x_2$, so if $x_2 = k$, then $x_1 = -k$, Thus, the eigenvector for $\lambda = 1$ is $\begin{bmatrix} -k \\ k \end{bmatrix}$. Check these answers using $(A - \lambda I)X = 0$:

$$\begin{bmatrix} 2 - 1 & 1 \\ 3 & 4 - 1 \end{bmatrix} \begin{bmatrix} -k \\ k \end{bmatrix} = \begin{bmatrix} 0 \\ 0 \end{bmatrix} \Rightarrow \begin{bmatrix} 1 & 1 \\ 3 & 3 \end{bmatrix} \begin{bmatrix} -k \\ k \end{bmatrix} = \begin{bmatrix} -k + k \\ 3(-k) + 3k \end{bmatrix} = \begin{bmatrix} 0 \\ 0 \end{bmatrix}$$

Now determine the eigenvector for $\lambda = 5$. Begin with $(A - \lambda I)X = 0$:

$$\begin{bmatrix} 2-5 & 1 \\ 3 & 4-5 \end{bmatrix} \begin{bmatrix} x_1 \\ x_2 \end{bmatrix} = \begin{bmatrix} 0 \\ 0 \end{bmatrix} \Rightarrow \begin{bmatrix} -3 & 1 \\ 3 & -1 \end{bmatrix} \begin{bmatrix} x_1 \\ x_2 \end{bmatrix} = \begin{bmatrix} 0 \\ 0 \end{bmatrix} \Rightarrow \begin{bmatrix} -3x_1 + x_2 \\ 3x_1 - x_2 \end{bmatrix} = \begin{bmatrix} 0 \\ 0 \end{bmatrix}$$

From the first equation we know $-3x_1 + x_2 = 0 \Rightarrow x_2 = 3x_1$, so if $x_2 = k$, then $x_1 = \frac{1}{3}k$, and the eigenvector for $\lambda = 5$ is $\begin{bmatrix} \frac{1}{3}k \\ k \end{bmatrix}$ (or $\begin{bmatrix} k \\ 3k \end{bmatrix}$ if you let $x_1 = k$). Use $(A - \lambda I)X = 0$ again to check these answers.

$$\begin{bmatrix} 2-5 & 1 \\ 3 & 4-5 \end{bmatrix} \begin{bmatrix} \frac{1}{3}k \\ k \end{bmatrix} = \begin{bmatrix} 0 \\ 0 \end{bmatrix} \Rightarrow \begin{bmatrix} -3 & 1 \\ 3 & -1 \end{bmatrix} \begin{bmatrix} \frac{1}{3}k \\ k \end{bmatrix} = \begin{bmatrix} -k+k \\ k-k \end{bmatrix} = \begin{bmatrix} 0 \\ 0 \end{bmatrix}$$

(Notice the trace of $A = 6$, which was the sum of the eigenvalues, and $|A| = 5$, which was the product of the eigenvalues.)

Problem 6.5.2. Find the eigenvalues and eigenvectors for $B = \begin{bmatrix} 2 & -2 & 3 \\ 1 & 1 & 1 \\ 1 & 3 & 1 \end{bmatrix}$. Begin by constructing $|B - \lambda I|$:

$$\begin{vmatrix} 2 & -2 & 3 \\ 1 & 1 & 1 \\ 1 & 3 & -1 \end{vmatrix} - \lambda \begin{vmatrix} 1 & 0 & 0 \\ 0 & 1 & 0 \\ 0 & 0 & 1 \end{vmatrix} = \begin{vmatrix} 2 & -2 & 3 \\ 1 & 1 & 1 \\ 1 & 3 & -1 \end{vmatrix} = \begin{vmatrix} \lambda & 0 & 0 \\ 0 & \lambda & 0 \\ 0 & 0 & \lambda \end{vmatrix} = \begin{vmatrix} 2-\lambda & -2 & 3 \\ 1 & 1-\lambda & 1 \\ 1 & 3 & -1-\lambda \end{vmatrix}$$

Now solve the equation $|B - \lambda I| = 0$:

$$|B - \lambda I| = 0 = \begin{vmatrix} 2-\lambda & -2 & 3 \\ 1 & 1-\lambda & 1 \\ 1 & 3 & -1-\lambda \end{vmatrix} = \cdots = \lambda^3 - 2\lambda^2 - 5\lambda + 6$$

There is nothing particularly new about calculating this determinant, so I have deleted the middle steps. We know the sum of the eigenvalues must equal 2, which is the trace of B. We also know the product of the eigenvalues must equal -6, which is $|B|$. Thus, it is a good guess that the eigenvalues will be some combination of 1, 2, and 3 (positive or negative). It should only take a bit of experimentation to find

$$0 = \lambda^3 - 2\lambda^2 - 5\lambda + 6 = (\lambda - 1)(\lambda^2 - \lambda - 6) + (\lambda - 1)(\lambda + 2)(\lambda - 3)$$
$$\therefore \lambda = -2, 1, 3$$

Now use $(\mathbf{B} - \lambda\mathbf{I})\mathbf{X} = \mathbf{0}$ to determine the eigenvector for $\lambda = -2$:

$$\begin{bmatrix} 2-(-2) & -2 & 3 \\ 1 & 1-(-2) & 1 \\ 1 & 3 & -1-(-2) \end{bmatrix}\begin{bmatrix} x_1 \\ x_2 \\ x_3 \end{bmatrix} = \begin{bmatrix} 4 & -2 & 3 \\ 1 & 3 & 1 \\ 1 & 3 & 1 \end{bmatrix}\begin{bmatrix} x_1 \\ x_2 \\ x_3 \end{bmatrix} = \begin{bmatrix} 4x_1 - 2x_2 + 3x_3 \\ x_1 + 3x_2 + x_3 \\ x_1 + 3x_2 + x_3 \end{bmatrix} = \begin{bmatrix} 0 \\ 0 \\ 0 \end{bmatrix}$$

The solution here is a bit more difficult. Manipulate the second equation to find that $x_1 + 3x_2 + x_3 = 0 \Rightarrow x_1 = -3x_2 - x_3$. Now substitute this value into the first equation:

$$0 = 4(-3x_2 - x_3) - 2x_2 + 3x_3 = -12x_2 - 4x_3 - 2x_2 + 3x_3 = -14x_2 - x_3$$
$$\Rightarrow x_3 = -14x_2$$

To avoid the use of fractions, let $x_2 = k$, then $x_3 = -14k$. We can now use these values to find that $x_1 = -3k - (-14k) = 11k$. Thus, the eigenvector for $\lambda = -2$ is $[11k \ k \ -14k]^\mathrm{T}$. To verify this answer, use $(\mathbf{B} - \lambda\mathbf{I})\mathbf{X} = \mathbf{0}$ as follows:

$$\begin{bmatrix} 2-(-2) & -2 & 3 \\ 1 & 1-(-2) & 1 \\ 1 & 3 & -1-(-2) \end{bmatrix}\begin{bmatrix} 11k \\ k \\ -14k \end{bmatrix} = \begin{bmatrix} 4 & -2 & 3 \\ 1 & 3 & 1 \\ 1 & 3 & 1 \end{bmatrix}\begin{bmatrix} 11k \\ k \\ -14k \end{bmatrix}\begin{bmatrix} 44k - 2k - 42k \\ 11k + 3k - 14k \\ 11k + 3k - 14k \end{bmatrix} = \begin{bmatrix} 0 \\ 0 \\ 0 \end{bmatrix}$$

Turning to the eigenvector for $\lambda = 1$, we find

$$\begin{bmatrix} 2-1 & -2 & 3 \\ 1 & 1-1 & 1 \\ 1 & 3 & -1-1 \end{bmatrix}\begin{bmatrix} x_1 \\ x_2 \\ x_3 \end{bmatrix} = \begin{bmatrix} 1 & -2 & 3 \\ 1 & 0 & 1 \\ 1 & 3 & -2 \end{bmatrix}\begin{bmatrix} x_1 \\ x_2 \\ x_3 \end{bmatrix} = \begin{bmatrix} x_1 - 2x_2 + 3x_3 \\ x_1 + 0 + x_3 \\ x_1 + 3x_2 - 2x_3 \end{bmatrix} = \begin{bmatrix} 0 \\ 0 \\ 0 \end{bmatrix}$$

From the second equation we can see that if $x_1 = k$, then $x_3 = -k$. Substitute these values into the first equation to find $k - 2x_2 - 3k = 0 \Rightarrow$ $x_2 = -k$. Thus, the eigenvector for $\lambda = 1$ is $[k \ -k \ -k]^T$.

Finally, to find the eigenvector for $\lambda = 3$, we begin as usual:

$$\begin{bmatrix} 2-3 & -2 & 3 \\ 1 & 1-3 & 1 \\ 1 & 3 & -1-3 \end{bmatrix} \begin{bmatrix} x_1 \\ x_2 \\ x_3 \end{bmatrix} = \begin{bmatrix} -1 & -2 & 3 \\ 1 & -2 & 1 \\ 1 & 3 & -4 \end{bmatrix} \begin{bmatrix} x_1 \\ x_2 \\ x_3 \end{bmatrix} = \begin{bmatrix} -x_1 - 2x_2 + 3x_3 \\ x_1 - 2x_2 + x_3 \\ x_1 + 3x_2 - 4x_3 \end{bmatrix} = \begin{bmatrix} 0 \\ 0 \\ 0 \end{bmatrix}$$

From the first equation we see that $x_1 = -2x_2 + 3x_3$. Substituting this value into the second equation we find

$$0 = (-2x_2 + 3x_3) - 2x_2 + x_3 = -4x_2 + 4x_3$$
$$\Rightarrow 4x_2 = 4x_3 \Rightarrow x_2 = x_3$$

If $x_2 = k$, then $x_3 = k$. Substituting these values into the previous equation yields $x_1 = -2k + 3k = k$. Thus, the eigenvector for $\lambda = 3$ is $[k \ k \ k]^T$. Verify for yourself that the eigenvectors for $\lambda = 1$ and $\lambda = 3$ are correct.

6.6 Multivariate Extrema and Matrix Algebra

Problem 6.6.1. Use a Hessian determinant to find the extreme points of $f(x_1, x_2, x_3) = x_1^2 + x_2^2 + x_3^2 + x_1x_2 + x_2x_3 - 3x_1 - 8$. Begin with the first-order condition:

$$f_{x_1} = 2x_1 + x_2 - 3 = 0$$
$$f_{x_2} = 2x_2 + x_1 + x_3 = 0$$
$$f_{x_3} = 2x_3 + x_2 = 0$$

Solve this system to find $x_1 = \dfrac{9}{4}$, $x_2 = \dfrac{-3}{2}$, and $x_3 = \dfrac{3}{4}$. Entering these values into $f(x_1, x_2, x_3)$ tells us $\left(\dfrac{9}{4}, \dfrac{-3}{2}, \dfrac{3}{4}, \dfrac{-91}{8}\right)$ is the critical point. Next find the Hessian determinant and the principal minors:

$$|\mathbf{H}| = \begin{vmatrix} f_{11} & f_{12} & f_{13} \\ f_{21} & f_{22} & f_{23} \\ f_{31} & f_{32} & f_{33} \end{vmatrix} = \begin{vmatrix} 2 & 1 & 0 \\ 1 & 2 & 1 \\ 0 & 1 & 2 \end{vmatrix}; \; |\mathbf{H}_1| = |2| = 2,$$

$$|\mathbf{H}_2| = \begin{vmatrix} 2 & 1 \\ 1 & 2 \end{vmatrix} = 3, \; |\mathbf{H}_3| = \begin{vmatrix} 2 & 1 & 0 \\ 1 & 2 & 1 \\ 0 & 1 & 2 \end{vmatrix} = 4$$

Recall that entries in $|\mathbf{H}|$ are second derivatives for the indicated variables. We know that from the second derivative test because $|\mathbf{H}_1| > 0$, $|\mathbf{H}_2| > 0$, and $|\mathbf{H}_3| > 0$, the critical point is a minimum.

Problem 6.6.2. Use a Hessian determinant to find the extreme points of $g(t_1, t_2, t_3) = 2t_1^2 + t_2^2 - 3t_3^2 + t_1t_2 - 2t_2t_3 - 29t_3 - 3$. Begin with the first-order condition:

$$g_{t_1} = 4t_1 + t_2 = 0$$
$$g_{t_2} = 2t_2 + t_1 - 2t_3 = 0$$
$$g_{t_3} = -6t_3 - 2t_2 - 29 = 0$$

Solve this system to find $t_1 = 1$, $t_2 = -4$, and $t_3 = \dfrac{-7}{2}$, which yields a critical point of $\left(1, 4, \dfrac{-7}{2}, \dfrac{191}{4}\right)$. Now find the Hessian determinant and principal minors:

$$|\mathbf{H}| = \begin{vmatrix} g_{11} & g_{12} & g_{13} \\ g_{21} & g_{22} & g_{23} \\ g_{31} & g_{32} & g_{33} \end{vmatrix} = \begin{vmatrix} 4 & 1 & 0 \\ 1 & 2 & -2 \\ 0 & -2 & -6 \end{vmatrix}; \; |\mathbf{H}_1| = |4| = 4,$$

$$|\mathbf{H}_2| = \begin{vmatrix} 4 & 1 \\ 1 & 2 \end{vmatrix} = 7, \; |\mathbf{H}_3| = \begin{vmatrix} 4 & 1 & 0 \\ 1 & 2 & -2 \\ 0 & -2 & -6 \end{vmatrix} = -58$$

Here the principal minors of the Hessian are not all greater than 0, nor do they alternate beginning with a negative sign (which would indicate a maximum), so we must examine a neighborhood around $\left(1, 4, \dfrac{-7}{2}, \dfrac{191}{4}\right)$. We do so by varying the value for each of the variables as follows:

t_1	t_2	t_3	$g(t_1, t_2, t_3)$
1	−4	$\frac{-7}{2}$	47.75
1	−4	−4	47
1	−4	−3	47
1	−5	$\frac{-7}{2}$	48.75
1	−3	$\frac{-7}{2}$	48.75
0	−4	$\frac{-7}{2}$	49.75
2	−4	$\frac{-7}{2}$	49.75

The first row of values represents the critical point. From these values we can see that the critical point is not an extremum. These values also point out the necessity of checking all three dimensions before concluding that the critical point is an extremum.

Problem 6.6.3. Use a bordered Hessian determinant to optimize $f(t_1, t_2, t_3) = 20t_1t_2t_3$, subject to the constraint $g(t_1, t_2, t_3) = 150 - 5t_1 - 10t_2 - 15t_3 = 0$. Begin by constructing $F(t_1, t_2, t_3, \lambda) = 20t_1t_2t_3 - \lambda(150 - 5t_1 - 10t_2 - 15t_3)$. To satisfy the first-order condition, take the first-order partials of F, set them equal to 0, and solve

$$F_{t_1} = 20t_2t_3 + 5\lambda = 0$$
$$F_{t_2} = 20t_1t_3 + 10\lambda = 0$$
$$F_{t_3} = 20t_1t_2 + 15\lambda = 0$$
$$F_\lambda = -150 + 5t_1 + 10t_2 + 15t_3 = 0$$

Solve this set of equations to find $t_1 = 10$, $t_2 = 5$, $t_3 = \frac{10}{3}$, and $\lambda = \frac{-200}{3}$ with a critical point of $\left(10, 5, \frac{10}{3}, \frac{10,000}{3}\right)$. Next construct the bordered Hessian determinant:

$$|\overline{\mathbf{H}}| = \begin{vmatrix} 0 & 5 & 10 & 15 \\ 5 & 0 & 20t_3 & 20t_2 \\ 10 & 20t_3 & 0 & 20t_1 \\ 15 & 20t_2 & 20t_1 & 0 \end{vmatrix} = \begin{vmatrix} 0 & 5 & 10 & 15 \\ 5 & 0 & \dfrac{200}{3} & 100 \\ 10 & \dfrac{200}{3} & 0 & 200 \\ 15 & 100 & 200 & 0 \end{vmatrix} = -3{,}000{,}000$$

Notice that this bordered Hessian determinant contains variables. Because we are testing whether the critical point is a maximum, I entered the values of the critical point for the appropriate variables ($20t_3 = 20 \times 5 = 100$, etc.). Now examine the principal minors of $|\overline{\mathbf{H}}|$. Because $m = 1$ (i.e., there was one constraint), begin with $|\overline{\mathbf{H}}_2|$:

$$|\overline{\mathbf{H}}_2| = \begin{vmatrix} 0 & 5 & 10 \\ 5 & 0 & \dfrac{200}{3} \\ 10 & \dfrac{200}{3} & 0 \end{vmatrix} = \dfrac{20{,}000}{3}, \text{ and } |\overline{\mathbf{H}}_3| = \begin{vmatrix} 0 & 5 & 10 & 15 \\ 5 & 0 & \dfrac{200}{3} & 100 \\ 10 & \dfrac{200}{3} & 0 & 200 \\ 15 & 100 & 200 & 0 \end{vmatrix} = -3{,}000{,}000$$

The signs alternate, beginning with $(-1)^{1+1} = 1$, which indicates that the critical point is a maximum. (*Note*: As we have seen previously, the same symbol can have different meanings depending on the context. In the previous section, λ was used to represent eigenvalues. In this section, λ refers to the Lagrange multipliers we first used in Section 4.2.2.)

Problem 6.6.4. A budget analyst in the Department of Defense determines that the utility function for three types of armored vehicles is $f(v_1, v_2, v_3) = \ln v_1 + \ln v_2 + \ln v_3$. If each of the first type of vehicle costs \$500,000, the second \$400,000 and the third \$1 million, and the total budget allocated for the three types is \$30 million, how many of each type will maximize the department's utility?

Begin by recognizing that the constraint is $g(v_1, v_2, v_3) = 300 - 5v_1 - 4v_2 - 10v_3 = 0$. (Dividing by 100,000 will make the calculations much easier and not change the results.) Now construct $F(v_1, v_2, v_3, \lambda) = \ln v_1$

+ $\ln v_2 + \ln v_3 - \lambda(300 - 5v_1 - 4v_2 - 10v_3)$. Next find the first-order partials of F, set them equal to 0, and solve

$$F_{v_1} = \frac{1}{v_1} + 5\lambda = 0$$

$$F_{v_2} = \frac{1}{v_2} + 4\lambda = 0$$

$$F_{v_3} = \frac{1}{v_3} + 10\lambda = 0$$

$$F_\lambda = -300 + 5v_1 + 4v_2 + 10v_3 = 0$$

Solve this system of equations to find $v_1 = 20$, $v_2 = 25$, $v_3 = 10$, and $\lambda = \frac{-1}{100}$. Next construct the bordered Hessian determinant:

$$|\overline{\mathbf{H}}| = \begin{vmatrix} 0 & 5 & 4 & 10 \\ 5 & \frac{-1}{v_1^2} & 0 & 0 \\ 4 & 0 & \frac{-1}{v_2^2} & 0 \\ 10 & 0 & 0 & \frac{-1}{v_3^2} \end{vmatrix} = \begin{vmatrix} 0 & 5 & 4 & 10 \\ 5 & \frac{-1}{400} & 0 & 0 \\ 4 & 0 & \frac{-1}{625} & 0 \\ 10 & 0 & 0 & \frac{-1}{100} \end{vmatrix} = \frac{-3}{2500}$$

Now examine the principal minors. Because $m = 1$, we begin with $|\overline{\mathbf{H}}_2|$:

$$|\overline{\mathbf{H}}_2| = \begin{vmatrix} 0 & 5 & 4 \\ 5 & \frac{-1}{400} & 0 \\ 4 & 0 & \frac{-1}{625} \end{vmatrix} = \frac{2}{25}, \text{ and } |\overline{\mathbf{H}}_3| = \begin{vmatrix} 0 & 5 & 4 & 10 \\ 5 & \frac{-1}{400} & 0 & 0 \\ 4 & 0 & \frac{-1}{625} & 0 \\ 10 & 0 & 0 & \frac{-1}{100} \end{vmatrix} = \frac{-3}{2500}$$

Again, the signs alternate, beginning with $(-1)^{1+1} = 1$, which indicates that the critical point (20, 25, 10, 8.52) is a maximum.

6.7 Homework Problems

1. Do the indicated operation, if possible, using these four matrices:

$$A = \begin{bmatrix} 3 & 2 \\ 4 & 1 \end{bmatrix}, B = \begin{bmatrix} 5 & 1 \\ 3 & 2 \end{bmatrix}, C = \begin{bmatrix} 0 & 2 \\ 1 & 2 \\ 4 & 1 \end{bmatrix}, D = \begin{bmatrix} 1 & 3 & 1 \\ 2 & 0 & 5 \end{bmatrix}$$

   a. **A + B**   e. **BA**
   b. **B − A**   f. **DB**
   c. **2A**   g. **BD**
   d. **AB**   h. **DC**

2. Calculate the determinants of the following matrices.

   a. $\begin{bmatrix} 6 & 2 \\ 3 & 8 \end{bmatrix}$

   d. $\begin{bmatrix} 3 & 2 & 3 \\ -1 & 4 & 1 \\ 1 & 3 & -2 \end{bmatrix}$

   b. $\begin{bmatrix} -2 & 4 \\ 1 & 3 \end{bmatrix}$

   e. $\begin{bmatrix} 1 & 2 & 3 & 1 \\ 0 & 3 & -1 & 4 \\ 2 & -1 & 1 & 3 \\ 3 & 3 & 2 & 1 \end{bmatrix}$

   c. $\begin{bmatrix} 1 & 2 & 1 \\ 0 & 3 & 1 \\ 2 & 3 & 2 \end{bmatrix}$

   f. $\begin{bmatrix} 3 & 2 & -1 & 0 \\ -2 & 4 & 1 & 1 \\ 1 & -2 & 2 & 3 \\ -1 & 1 & 2 & 4 \end{bmatrix}$

3. Find the inverses for the matrices in the previous problem.

4. Solve the following systems of equations using Cramer's rule.

   a. $3x - 4y + z = 6$
      $x + y - 2z = -5$
      $-2x + 3y + 3z = 11$

   b. $r + s - t = 2$
      $-r + 3s - 4t = 12$
      $3r - s + 5t = -2$

5. Find the eigenvalues and eigenvectors for the following matrices.

   a. $\begin{bmatrix} 4 & 3 \\ 2 & 5 \end{bmatrix}$

   c. $\begin{bmatrix} 2 & -1 & 2 \\ 0 & 1 & 2 \\ 2 & -2 & 3 \end{bmatrix}$

   b. $\begin{bmatrix} 6 & 2 \\ 4 & 8 \end{bmatrix}$

   d. $\begin{bmatrix} 4 & 1 & -1 \\ 1 & 2 & -1 \\ 1 & -1 & 2 \end{bmatrix}$

6. Find the critical points of the following functions and use a Hessian determinant to determine whether they are maxima or minima.

   a. $f(r, s, t) = r^2 + 8s^2 + t^2 - rt + 4$

   b. $f(x_1, x_2, x_3) = -x_2^3 - 3x_1^2 - x_3^2 + 3x_1x_2 + 2x_3 + 1$

7. Minimize the function in problem 6a subject to the constraint $g(r, s, t) = 33 - r - s - t$. Use a bordered Hessian determinant to verify that the point is a minimum.

8. The administrator of a social services agency has a $2 million budget to fund three programs. The first program can be funded in increments of $160,000, the second in increments of $50,000, and the third in increments of $200,000. If the administrator has determined that the benefit function for the three programs is $f(p_1, p_2, p_3) = 2 \ln p_1 + \ln p_2 + 2 \ln p_3$, how many increments of each program should be funded to maximize benefits? Use a bordered Hessian determinant to verify that the critical point is a maximum.

APPENDIX: ANSWERS TO HOMEWORK PROBLEMS

Chapter 1. 1. (a) $\frac{1}{16}$ (b) .001 (c) a^4b^2c 2. (a) 3 (b) 4 (c) 5 (d) m (e) 4/3 (f) $\frac{1}{15}$ 3. (a) function, see Figure A.1a (b) function, see Figure A.1b (c) not a function, see Figure A.1c 4. (a) $y = 2x$ (b) $y = \frac{3}{4}x - \frac{1}{4}$ 5. (a) (4, 2) (b) (0, -1) 6. (a) $x^2 - 6x + 8$ (b) $6x^2 - 2x - 28$ (c) $-3x^2 - 2x + 5$ (d) $x^2 - x - 12$ 7. (a) $(x + 1)(x + 2)$ (b) $3(x + 1)(x + 3)$ (c) $(-1)(x - 5)(x - 1)$ (d) $x(x + 1)(x + 1)$ 8. (a) $x = \pm\sqrt{5}$ (b) $x = \frac{1}{2}, -4$ (c) $x = 0$ 9. (a) $S = \{(1, 6), (2, 5), (3, 4), (4, 3), (5, 2), (6, 1)\}$ (b) $S = \varnothing$ 10. (a) $S = A \cap B \cap C'$ (b) A (c) $(A \cup C) \cap B'$ (d) $(B \cap D) \cap (A \cup C)'$ 11. (a-b) see Figure A.2 12. (a) 17,160 (b) 930 13. (a) 120 (b) 421,200 (c) $23 \times 22 \times 25 \times 24 = 303,600$ 14. 40,320; $4! \times 3! \times 2! \times 2! \times 1! = 576$ 15. (a) 84 (b) 330 (c) 4,368

Chapter 2. 1. (a) limit exists (b) limit exists (c) a not in domain of f, limit does not exist (d) limit exists 2. (a) no limit, goes to $-\infty$ from the left and ∞ from the right (b) ∞ (c) 0 (d) $\frac{9}{2}$, first multiply the numerator 3. (a) 7 (b) 1 (c) 14 (d) 16 (e) 3 (f) –4 (g) no limit, goes to ∞ from the

98

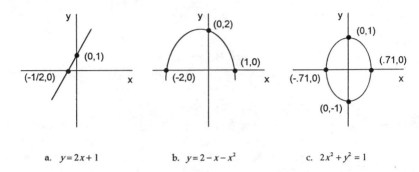

a. $y = 2x + 1$ b. $y = 2 - x - x^2$ c. $2x^2 + y^2 = 1$

Figure A.1. Solution Graphs for Homework Problem 1.3

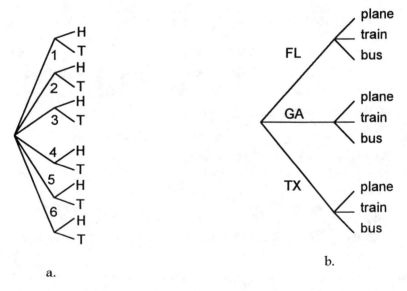

a.

b.

Figure A.2. Solution Tree Diagrams for Homework Problem 1.11

left and $-\infty$ from the right (h) $\frac{5}{3}$ (i) 5 (j) $\frac{1}{6}$, for the positive square root
4. (a) everywhere except $x = 2/3$ (b) everywhere except $x = 2$ (c) continuous for $x \geq -3$, except at $x = 0$

Chapter 3. 1. 11 2. (a) $f'(x) = 0$ (b) $g'(x) = -7$ (c) $f'(t) = 12t^2 - 7$ (d) $f'(x) = 4x^3 + 2x^{-3}$ (e) $g'(x) = -3x^{-2} - 5x^{-6}$ (f) $h'(x) = 7x^{-2/9} + \frac{3}{2}x^{-7/4} - 16x^{-4.2}$
3. (a) $g'(x) = 6x^2 - 10x - 18$ (b) $f'(t) = 14t^6 + 6t^2 + 12t - 12t^{-3}$ (c) $h'(x) = 24x^3 + 141x^2 - 82x - 68$ (d) $h'(t) = -28t^{1/6} - t^{-7/6} - 18$ 4. (a) $g'(t) = 21(t + 7)^{-2}$ (b) $f'(t) = -2t^{-3} + 16t^{-5}$ (c) $h'(x) = \dfrac{-x^4 - 14x^3 + 12x^2 - 2x - 7}{(x^3 - 1)^2}$
(d) $g'(x) = 4 + \dfrac{x^4 + 3x^2}{(1 + x^2)^2}$ 5. (a) $h'(x) = 6x(x^2 - 9)^2$ (b) $h'(t) = 72t^7 + 30t^4$
$- 168t^3 + 2t - 14$ (c) $g'(t) = (3t^5 + 4)(t^6 + 8t)^{-1/2}$ (d) $f'(x) = \left(\dfrac{-3}{4}\right)(6 + 2x - 6x^2)(6x + x^2 - 2x^3)^{-7/4}$ 6. (a) $f'(x) = 2e^{2x}$ (b) $g'(x) = 3x^2e^{x^3}$ (c) $h'(x) = (8x^3 - 2x)e^{2x^4 - x^2}$ (d) $f'(t) = (90t^4 + 12t)e^{5t^3}$ (e) $g'(t) = (14t^3 + 10t)e^{t^2}$ (f) $h'(t) = \dfrac{(25t^5 + 5t^4 - 10)\,e^{t^5}}{(1 + 5t)^3}$ 7. (a) $h'(x) = \dfrac{3}{x}$ (b) $g'(x) = 20x^4(\ln x) + 4x^4 - 3x^{-1}$ (c) $f'(x) = 3x^{-1}(\ln x)^2$ (d) $f'(t) = \dfrac{2}{x(1 - x^4)} + \dfrac{4x^3 \ln x^2}{(1 - x^4)^2}$ 8. (a) $f'(x) = 9x^2$;
$f''(x) = 18x$ (b) $g'(x) = -(x + 1)^{-2}$; $g''(x) = 2(x + 1)^{-3}$ (c) $g'(t) = (5 + 2t)e^{2t}$;
$g''(t) = (12 + 4t)e^{2t}$ (d) $h'(t) = 2te^{t^2 - 1}$; $h''(t) = (4t^2 + 2)e^{t^2 - 1}$
9. (a) $h'(x) = 8x$; $h'(2) = 16$ (b) $h'(t) = 16t^3 - 8t$; $h'(1) = 8$ (c) $g'(x) = \dfrac{x^4 + 2x^3 + 15x^2 + 2x + 1}{(1 - x^3)^2}$; $g'(-1) = \dfrac{13}{4}$ (d) $g'(t) = 3t^2e^{t^3}$; $g'(1) = 3e \approx 8.15$
10. (a) maximum at $(0, 3)$ (b) maximum at $(-1, 1)$; minimum at $\left(\dfrac{1}{3}, \dfrac{-5}{27}\right)$, (c) critical point $(0, 0)$ an inflection point (d) minimum at $(0, 1)$ 11. minimum at $(1, -2)$; global maximum at the endpoint $(2, 2)$; relative maximum at the endpoint $(0, 0)$ 12. (a) $S'(x) = 100x + 100$ (b) $S'(4) = 500$, $S(5) - S(4) = 550$ (c) assumes rate of subscriptions will always be increasing 13. 5 $(g(p) = 3000p - 40p^3)$ 14. 36" by 48"

Chapter 4. 1. (a) $f_x = 12x^3$; $f_y = -9$ (b) $f_x = 21x^2y^2z^7$; $f_y = 14x^3yz^7$; $f_z = 49x^3y^2z^6$ (c) $f_s = 24s^2t + 10st^2 - t^3$; $f_t = 8s^3 + 10s^2t - 3st^2$ (d) $h_s = 6 - 6t$;

$h_t = 16t^3 - 12t^2 - 6s$ (e) $f_x = f_y = -(x + y)^{-2}$, factor the denominator and reduce (f) $f_r = 24rs^4(4r^2s^4 - 9)^2$; $f_s = 48r^2s^3(4r^2s^4 - 9)^2$ (g) $g_x = 2ye^{2xy}$; $g_y = 2xe^{2xy}$ (h) $f_x = 2xe^{x^2-9y^2}$; $f_y = -18ye^{x^2-9y^2}$ **2.** (a) $f_x = 6x^2$; $f_y = -14y$; $f_{xx} = 12x$; $f_{yy} = -14$; $f_{xy} = f_{yx} = 0$ (b) $f_s = 27s^2t - 3t^2$; $f_t = 9s^3 - 6st + 12t^2$; $f_{ss} = 54st$; $f_{tt} = -6s + 24t$; $f_{st} = f_{ts} = 27s^2 - 6t$ (c) $f_u = 4u^3 - 2uv - v^3$; $f_v = -u^2 - 3uv^2$; $f_{uu} = 12u^2 - 2v$; $f_{vv} = -6uv$; $f_{uv} = f_{vu} = -2u - 3v^2$ (d) $f_x = 7x^6y^2 - 6y$; $f_y = 2x^7y - 6x$; $f_{xx} = 42x^5y^2$; $f_{yy} = 2x^7$; $f_{xy} = f_{yx} = 14x^6y - 6$ **3.** (a) minimum at $(-3, 3, -5)$ (b) minimum at $(-8, 15, -107)$ (c) maximum at $(-1, -3, 9)$ (d) critical point $(-2, 1, 3)$ not an extremum **4.** minimum at $(2, -1, 19)$ **5.** maximum at $x = y = z = N/3$ **6.** local booklet = 53 cents; national booklet = 55 cents **7.** $h = w = l$ **8.** $\frac{2}{3}h = w = l$

Chapter 5. **1.** (a) $4x + c$ (b) $4x^2 + c$ (c) $\frac{1}{8}u^8 + c$ (d) $-3z^{-2} + c$ (e) $2x^3 + x^2 - 9x + c$ (f) $\frac{3}{7}x^{7/3} + 3x^{1/3} + c$ (g) $\frac{8}{5}t^5 - 2t^4 + \frac{2}{3}t^3 + c$ (h) $\frac{-1}{3}z^{-1} + c$, factor the denominator and reduce (i) $7e^x + c$ (j) $-2x^{-4} + 4e^x + c$ **2.** (a) $\frac{1}{28}(7x - 9)^4 + c$ (b) $\frac{1}{4}e^{4t} + c$ (c) $\frac{1}{5}(2z^2 - 5)^5 + c$ (d) $\frac{1}{90}(6t^3 - 8)^5 + c$ (e) $e^{3x^3} + c$ (f) $\frac{-1}{24}(3x^{-2} + 2)^4 + c$ **3.** (a) $\frac{2}{3}e^{3z}\left(z - \frac{1}{3}\right) + c$ (b) $e^{-x}(-x - 2) + c$, multiply before integrating (c) $\frac{2}{3}x(x + 6)^{3/2} - \frac{4}{15}(x + 7)^{5/2} + c$ (d) $e^{3t}\left(\frac{1}{3}t^2 - \frac{2}{9}t + \frac{2}{27}\right) + c$ **4.** (a) $\frac{63}{2}$ (b) -6 (c) $\frac{-3487}{96}$ (d) $\frac{20}{3}$ (e) $\frac{1}{12}$ (f) $\frac{48(\ln 4) - 28}{9} \approx 4.28$ **5.** 5003; 5 **6.** 40; $120{,}005 - 5e^8 \approx 105{,}100$ **7.** 15 years; \$1,687.50; 10.29 or 11 complete years

Chapter 6. **1.** (a) $\begin{bmatrix} 8 & 3 \\ 7 & 3 \end{bmatrix}$ (b) $\begin{bmatrix} 2 & -1 \\ -1 & 1 \end{bmatrix}$ (c) $\begin{bmatrix} 6 & 4 \\ 8 & 2 \end{bmatrix}$ (d) $\begin{bmatrix} 21 & 7 \\ 23 & 6 \end{bmatrix}$ (e) $\begin{bmatrix} 19 & 11 \\ 17 & 8 \end{bmatrix}$

(f) **D** and **B** are nonconformable (g) $\begin{bmatrix} 7 & 15 & 10 \\ 7 & 9 & 13 \end{bmatrix}$ (h) $\begin{bmatrix} 7 & 9 \\ 20 & 9 \end{bmatrix}$ **2.** (a) 42 (b) -10 (c) 1 (d) -56 (e) 144 (f) 60 **3.** (a) $\frac{1}{42}\begin{bmatrix} 8 & -2 \\ -3 & 6 \end{bmatrix}$ (b) $\frac{1}{10}\begin{bmatrix} -3 & 4 \\ 1 & 2 \end{bmatrix}$ (c)

$$\begin{bmatrix} 3 & -1 & -1 \\ 2 & 0 & -1 \\ -6 & 1 & 3 \end{bmatrix} \text{(d)} \frac{1}{56}\begin{bmatrix} 11 & -13 & 10 \\ 1 & 9 & 6 \\ 7 & 7 & -14 \end{bmatrix} \text{(e)} \frac{1}{144}\begin{bmatrix} -45 & -15 & 18 & 51 \\ 3 & 17 & -30 & 19 \\ 57 & -13 & 6 & -23 \\ 12 & 20 & 24 & -20 \end{bmatrix}$$

$$\text{(f)} \frac{1}{60}\begin{bmatrix} 13 & 7 & 19 & -16 \\ 7 & 13 & 1 & -4 \\ -7 & 47 & 59 & -56 \\ 5 & -25 & -25 & 40 \end{bmatrix}$$ **4.** (a) $x = 2, y = 1, z = 4$ (b) $r = -2, s = 6, t = 2$

5. (a) eigenvalues: 2 and 7; eigenvectors: $[-3k \; 2k]^T$ and $[k \; k]^T$ (b) eigenvalues: 4 and 10; eigenvectors: $[-k \; k]^T$ and $[k \; 2k]^T$ (c) eigenvalues: 1, 2, and 3; eigenvectors: $[k \; k \; 0]^T$, $[3k \; 4k \; 2k]^T$, and $[k \; k \; k]^T$ (d) eigenvalues: 1, 3, and 4; eigenvectors: $[0 \; k \; k]^T$, $[k \; 0 \; k]^T$, and $[3k \; k \; k]^T$ **6.** (a) minimum at $(0, 0, 0, 4)$ (b) two critical points: maximum at $\left(\frac{1}{4}, \frac{1}{2}, 1, \frac{33}{16}\right)$, test fails at $(0, 0, 1, 2)$—check of neighborhood shows no extremum **7.** minimum at $(16, 1, 16, 268)$ **8.** maximum at $(5, 8, 4, 8.07)$

ABOUT THE AUTHOR

TIMOTHY M. HAGLE is Associate Professor in the Department of Political Science at the University of Iowa. He received his B.S. in Mathematics, B.A. in Communications (both with teaching certificate), and M.A. and Ph.D. in Political Science from Michigan State University. He also received his J.D. from Thomas M. Cooley Law School. He has taught mathematics to high school and undergraduate students and currently teaches the material in this monograph to first-year graduate students. He has published several articles in the area of judicial politics and behavior. His current research interests are in U.S. Supreme Court decision making and behavior.